Pfeiffer's Introduction to Biodynamics

Pfeiffer's
Introduction to
Biodynamics

Ehrenfried E. Pfeiffer

Floris
Books

First published as articles in *Bio-Dynamics,* the journal of
the Biodynamic Association, Inc.
'Rudolf Steiner's Contribution to Agriculture' appeared in
spring/summer 1956 as 'Rudolf Steiner's Impulse to Agriculture'
'The Biodynamic Preparations' appeared in summer 1948 as
'The Biodynamic Method, What it is and What it is Not'
'Questions and Answers on Biodynamics' appeared in autumn 1956 as
'Biodynamics, A Short, Practical Introduction'
'Ehrenfried Pfeiffer Pioneer in Agriculture and Natural Sciences'
by Herbert H. Koepf first published by the Biodynamic Association, Inc in 1991

This combined edition first published by Floris Books, Edinburgh
in cooperation with the Biodynamic Association, Inc in 2011
www.biodynamics.com
Third printing 2021

British Library CIP Data available
ISBN 978-086315-848-3
Printed in Great Britain by Bell & Bain, Ltd

Floris Books supports sustainable forest management by
printing this book on materials made from wood that
comes from responsible sources and reclaimed material

Contents

All measurements in this book are US and metric, not imperial. The US gallon at 3.8 litres is smaller than the imperial gallon at 4.5 litres. The ton is a short ton.

Rudolf Steiner's Contribution to Agriculture

The biodynamic farming and gardening method has grown and developed since 1922, based on advice and instruction given by Rudolf Steiner (1861–1925), the founder of anthroposophy.

The name "biodynamic" refers to a working with the energies which create and maintain life. This is what was meant in the name given it by the first group of farmers who put this new method to field use as well as practical tests. They decided to call it "biodynamic," a term derives from two Greek words, *bios* (life) and *dynamis* (energy).

In the years 1922–23, the problem of the increasing degeneration of seed stocks and of many cultivated plants brought a number of farmers, including Ernst Stegemann, to Rudolf Steiner for advice.

"What is to be done to stop the progressive ruin of seed quality and nutritional values?" It was in this form that the question was presented.

Among the observed facts to which reference was made, the following were perhaps among the most important: alfalfa could, in earlier times, be grown and harvested for as long as thirty years in the same field, then only for nine years, then for seven; at the time of the interview one was quite happy to maintain a stand for four to five years. Formerly a farmer could use his own rye, wheat, oats, barley, for many years as a seed stock. Now one always had to use new varieties after short periods of time. There were innumerable varieties which, after a few years, would disappear without trace. Different questions, related to the increase of animal diseases, especially those connected with sterility and the growing incidence of hoof and mouth disease, had brought other people, such as Dr Jospeh Werr and Dr Eugen Kolisko, and others working with the newly-developing Weleda drug company, to come to Dr Steiner with these problems.

A third impetus was given by Carl, Count von Keyserlingk, while questions having to do with the etheric aspect of plants, and the formative forces in general were presented by Dr Guenther Wachsmuth and myself. Answering such a query, about plant diseases, Dr Steiner explained to me that actually it was not the plant itself that was in the first place diseased, "since it is formed out of the healthy etheric." Rather it was the environment, especially the soil. One would need to look for the cause of the so-called plant diseases in the condition of the soil and of the total environment. Indications with regard to the inner attitude of those who farm, and concerning the steps toward the development of new cultivated plants were given especially to Ernst Stegemann in the years preceding the coming biodynamic movement.

The first biodynamic preparations

In 1923 Dr Steiner for the first time gave the procedures for making the biodynamic fertilizer preparations. And he did so without adding any explanations — only as recipes: "If you do this and that." Dr Wachsmuth and I then made the first preparation 500. It was buried in the garden of the Sonnenhof [home for children in need of special education] at Arlesheim, Switzerland.

Then, in the early summer of 1924 there came the memorable day when, in the presence of Dr Steiner, Dr Wegmann, Dr Wachsmuth and myself, as well as a number of other collaborators, this first preparation 500 was to be dug up again. It was a sunny afternoon, and we began to dig at the spot where we believed according to our memory — the memory supported by a number of landmarks — that we would find the preparation. The digging went on and on. The reader can conceive how we were sweating, not just from the exertion of the digging but because we were wasting Dr Steiner's valuable time. He also became impatient and got ready to leave, with the remark that he had to be back in his studio at 5 pm. At this moment the spade struck the first cowhorn.

Dr Steiner turned back and asked for a bucket of water and showed, then, how the content of the horn should be sprinkled into the water and stirred. My walking stick was the handiest available, and so it was used for the stirring. Dr Steiner stressed above all energetic stirring, the formation of vortices, and the quick reversals of direction which reveal the vortex-forming action of this energetic stirring. Nothing was said on the subject of manual stirring or of stirring with a bundle of birch twigs. There followed some brief advice as to how the stirred preparation was to be sprayed out and (illustrated by a motion of his hands) for how big an area the quantity at hand was to be used. And with that there came to an end the memorable event that stood at the birth-hour of a world-embracing agricultural movement.

What struck me and still gives much food for thought, is this development of things from one step to the next; from which one can see how objectively Dr Steiner worked. He never proceeded from a previously-held, abstract aim of teaching, but out of the given facts at hand.

Many years previously Dr Ludwig Noll had made known various suggestions of Dr Steiner's for the growing of medicinal herbs with strengthened effects of metallic substances or of silica. Dr Steiner said to me that this material was valid only for medicinal plants, that one should under no circumstances use the metallic additions with the fertilizer preparations used for food plants. He pointed to the fundamental difference between medicinal and nutritive plants, which extends so far that a plant grown for its medicinal values could lose its effectiveness if it were heavily fertilized like a food plant. On the other hand, the use of the metals on food plants would even have the effect of injuring health. This, as we grasped it, referred in particular to the application of metallic seed disinfectants and of pesticides (copper, mercury, lead, arsenic), as well as certain stone meal preparations.

At this time Count Keyserlingk was trying to persuade Dr Steiner to give an agricultural course. Dr Steiner was heavily overloaded with work, traveling, lecturing, and postponed the decision from week to week. Count Keyserlingk sent his nephew, Alexander von Keyserlingk to Dornach, and this young man explained that he would simply sit on Dr Steiner's doorstep and not go away until he had his firm commitment to presenting a course. This assurance was then given.

The Agricultural Course was held from June 7 to 16, 1924 in the hospitable home of the Count and Countess Keyserlingk. Associated with it were also a number of discussions and anthroposophical lectures in Breslau, among which was the noted Address to Youth. I did not have the privilege of taking part in the Agricultural Course, since Dr Steiner had requested that I help care for a severely ill friend.

"I will write you what goes on in the Course," he said by way of consolation. The overload of work resulted in the letter never being written, which was well understood although regretted. The general situation, however, was discussed with Dr Steiner after his return.

To the question as to whether one should begin with setting up experiments by way of introducing the new method, Dr Steiner answered, "The most important thing now is that the blessings of the preparations be carried to as large areas of land as possible, over the whole earth, for the healing of the earth and in order to improve the nutriment value of the field crops as widely as this can be done. That should be the aim. The experiments can then still be made later."

Dr Steiner apparently thought that the suggested ways of working ought to be given immediate practical application.

It is against this background that one must see the whole Agricultural Course. Guidance was given for understanding and making practical use of forces which lead the spiritual forces that are also often called the cosmic forces into the plant world again.

In the course of this discussion about the measures to be put into practice, it was also stated that the effects of the preparations and of the method "are for all, for all farmers." That is, they were not the privileged possession of a small, chosen group. This is all the more to be stressed since the only ones admitted to the Agricultural Course were farmers, gardeners and scientists who had a background in the practicalities of their work as well as an anthroposophical background. The latter is necessary in order to understand and evaluate what Rudolf Steiner explains; the biodynamic method, however, can be employed by every farmer. This must be stressed, since later, many got the idea that one could not work biodynamically if one were not an anthroposophist. On the other hand, the fact that knowledge of the biodynamic method gradually leads those using it to another world picture, that they learn particularly to judge the biological, that is, life processes and interconnections in a different way to that of the materialistic, chemically inclined farmer, is of course something obvious. So too, will they bring to the play of forces in nature a greater degree of interest and conscious awareness. One must learn to understand that there is a difference between mere application of the method and creative collaboration. A working together in the development of practical applications with the spiritual-cultural center, the Natural Science Section at the Goetheanum in Dornach, Switzerland, was especially advised. From there the spiritually fructifying, creative element was to go out, while the others brought to it both the questions and the practical carrying out of the principles.

The name "biodynamic" agriculture was not given by Rudolf Steiner but arose from the circle of those who at the start concerned themselves with the practical application of this new direction of thought. In the Agricultural Course (in which about 60 persons took part), Rudolf Steiner had presented the basic, new directions of thought concerning the connections of the earth, of the soil, with the formative forces of the etheric, the astral and the ego-activity in nature. He had especially shown how the health of the soil and of the plant and animal world depends on nature being again brought into connection with the cosmic, creative and formative forces.

To the consciousness of these connections he added certain practical measures (in the so-called fertilizer preparations). "The point now," he said to me on one occasion, "is to carry this out in practical application, to 'translate' it."

How close to Rudolf Steiner's heart lay the idea of the collaboration of the School of Spiritual Science with practical life was something that came to expression in a conversation in another connection. At that time Dr Steiner expressed the idea that a teacher at this school ought only to work there a certain number of years (he mentioned three) and then should work in a practical field somewhere else, and thus, through constant alternation, the connection with real life and its demands would never be lost.

The circle of those who became inspired by the Agricultural Course and joined in the task from the practical farming side as well as from science, was in a state of constant growth. One need only mention such names as those of Guenther Wachsmuth, Count Keyserlingk, Ernst Stegemann, Erhard Bartsch, Franz Dreidax, Immanuel Voegele, Max Karl Schwarz, Nicolaus Remer, Franz Rulni, Ernst Jakobi, Otto Eckstein, Hans Heinze and many others who joined the group as time went on. Also there was Dr Joseph Werr, as the first veterinary. Out of the working together of those engaged in its practical application with the Natural Science Section at the Goetheanum there arose, then, the biodynamic movement. It soon reached out to Austria, Switzerland, Italy, England, France, the Scandinavian countries, over to the U.S.A., and today it had collaborators in all parts of the world.

Justis von Liebig and beyond

At the time of the Agricultural Course the biodynamic direction of thought, and agricultural chemistry, stood opposed. The latter is essentially based on the views of Justus von Liebig, and it sought the sole explanation of the nutritive requirements of the plants in explaining the taking up of substance by the plant from the soil. Out of this arose the one-sided chemical fertilizer teaching of the Nitrogen-Phosphate-Potassium-Lime (NKP) requirements of the cultivated plants. This fertilizer teaching still rules the ortho-dox, scientifically orientated agriculture of today. But this teach-ing does not do completele justice to von Liebig. He himself had expressed doubts that the exact application of the NPK teaching is suitable for all soils. Deficiency phenomena appeared in greater intensity on humus-poor soils than on soils rich in humus.

However, the following quotation shows, in a far deeper sense, that von Liebig was not at all the hard-boiled materialist into which his followers turned him:

> The inorganic forces always create only that which
> is inorganic. Through the effects of a higher force
> that works in the living body, a power of which the
> inorganic forces are the servants, there comes into
> being the organic, characteristically formed substance,
> which is different from crystal and endowed with vital
> characteristics ... The cosmic conditions necessary for
> the plant being are warmth and sunlight.

"A higher force that works in the living body," "the cosmic con-ditions necessary." The answer to this was given by Rudolf Steiner. He resolved the problem posed by von Liebig, just through the fact that he did not remain standing at the purely substantial side of plant life but, with spiritual courage and with no preconcep-tions, took the next step.

An interesting situation then developed. The adherents of the purely material theory, who believed they had to reject the progressive thought indicated by Rudolf Steiner, are today forced by the facts of research in soil biology to go at least one step further. Things already recognized in 1924–34 in the circles of those with biodynamic knowledge have today become common knowledge. These include: the significance of the soil life, the soil being a living organism, the role of humus and the necessity of maintaining humus under all circumstances, and, where it is lacking, of producing it. To the undeniable nutritive relationship between plant and soil there has today been added knowledge of the biological, organic laws. One can go so far as to say that the biological part of the biodynamic method has become common knowledge. Indeed perhaps the mark has been overshot with it all. Despite recognition of the importance of the biological conditions of ecological relationships and interweaving of the plant world, questions of soil structure, the biological control of pests, the progress in the field of humus economy, all of this still gives no answer to the question of energy or its source, that is to say, of the cosmic conditions required by the plant. In a definite way the "biological" direction of thought has been taken up, although materialized. The dynamic side still waits understanding to which Rudolf Steiner pointed the way.

Since 1924 a number of works have been published that appear as an initial groping by science in this direction. We refer to the publications having to do with growth-regulating factors, the so-called growth-substances, enzymes, hormones, vitamins, trace elements and biocatalysts. But these fumbling searches always remain in the realm of substances. One has, indeed, come so far that effects of substances in dilutions of one to one million, even one to 100 million are no longer assigned to the realm of the unbelievable or fantastic and met with skeptical laughter such as was initially encountered by the rules for application of the biodynamic fertilizer preparations. These, with a dilution of between one to ten and one to a hundred million are still quite conceivable — conceivable, that is, in line with the present stage of knowledge. In the knowledge of photosynthesis, the building up of substance in the living plant cell, one has already broached the problem of the influence of energy (sun, light, warmth, the moon). This field of study deals with the transformation of cosmic sources of energy into chemically-substantially effective energy and states. One can read with interest in this connection the following, by W.R. Williams, a Member of the Academy of Science of the USSR. They are excerpts from his *Principles of Agriculture* published in 1952:

> The task of agriculture is to transform the varying sun energy, the energy of the light, into the inner force of foodstuffs for human consumption. The light is the basic raw material of the agricultural industry. Light and warmth are the necessary conditions of plant life. Light is the raw material out of which the agricultural products are made and warmth is the power which drives this mechanism of the plant. The dynamic energy of the sun radiation is transformed by the green plant into the material form of organic substance. So our first concrete task is the unending creation of organic substance in order to store up the inner energy for human life.

> One can divide the four essential factors into two groups according to their origins: light and warmth as cosmic factors, water and plant nutriments as terrestrial factors. The first group stems from inter-planetary space ...

> The cosmic factors work directly on the plant while the terrestrial factors only work through an intermediary, through substance.

The author of this work, which appeared in Russian, designates knowledge of the working together of the cosmic and terrestrial factors as the first objective of agricultural science, and knowledge of organic substance (humus) as the second.

In 1924 Rudolf Steiner pointed to the necessity of again consciously bringing the cosmic forces, both direct and indirect, into the growth processes, that is, to rescue knowledge of the being of the plant from its isolation in the purely terrestrial substance. Only thus would it be possible again to make effective the health-building and upbuilding forces that work against the degenerative processes. "By the middle of the century," he said in the course of his advice to me, "spiritual-scientific knowledge must have become life practice in order to prevent unspeakable injury to the health of nature and mankind."

Research methods

The methods of research was first to demonstrate the formative forces, and to find something by which the formative forces could be tested and demonstrated, then to show the weak points of the materialistic concept and to refute the results of materialistic research by means of its own experimental methods; that is, to use exact analytical methods in the realm of substance and develop them further. The idea throughout was to work quantitatively, not just qualitatively. Dr Steiner gave suggestions how the formative forces could be demonstrated, which I later developed in the crystallization method. And during my college studies, I had to see Dr Steiner each semester to present my plan of studies, and was advised in detail by Dr Steiner about the choice of subjects. In this connection Rudolf Steiner advised taking two, indeed three chemical laboratory courses — analytical, physical and botanical — at the same time. To the objection that this would simply not be possible in terms of hours per day, he said only, "Oh, you will be able to overcome that problem."

The emphasis was constantly on practical and laboratory work, not on clever theorizing.

In the course of the ten year period of work that arose out of this, these suggestions stood before my soul inducing me not only to work in the laboratory but also to apply the reality of the knowledge gained to the management of agricultural enterprises. This was done not only for the purpose of biodynamics but also from an economic point of view. "Unless one works economically, that is, in a way that brings in profit, the whole thing simply won't work," was Dr Steiner's instruction at the very beginning. In addition to the scientific studies Steiner also suggested I attend classes in political economy.

From these and many other suggestions it now became entirely clear what was to be done to introduce the biodynamic method. There was the large circle of the practical farmers. It was for them to carry out the application of the method on the farms. The best conditions for the effectiveness of the preparations — humus-building instead of humus-destroying crop rotations — had to be developed. The basic points of view for animal and plant breeding evolved themselves. It took years until the basic ideas were "translated" into practice. All of this was tested by the hard experiences of daily life, until the complete picture was there — a teachable and learnable method in which every farmer can share to their advantage. Questions that had to be answered in practice included those of soil cultivation, crop successions, the treatment of manure and compost, turning the heaps at the right time, animal husbandry and breeding, the care of fruit trees, and many others.

Then followed the discussions with representatives of agricultural science. For this, it was necessary to accumulate facts, observed data from laboratory research and field tests. Now my technical and quantitative-chemical education came in useful. It is in this sphere, where perhaps the gaps or weaknesses of the chemical theory of soils and nutrition showed themselves most clearly, that one today — after more than thirty years — sees the possibility of building a bridge between the perception of cosmic forces and an exact science.

The first sign of a breakthrough in the stagnant body of opinion perhaps showed itself in the discoveries around the concept of trace elements. In 1924 Rudolf Steiner pointed to these fine amounts of matter that are distributed in the atmosphere or otherwise, mentioning especially that they contribute very much to the healthy development of the plant world. The question still remained open whether these fine amounts of matter were taken up through the roots out of the soil or out of the atmosphere by the leaves and other organs of the plant. At the beginning of this thirty-year period it was determined by spectrum analysis that almost all elements are present in the atmosphere in dilutions from 10^8 to 10^9. The question about whether these trace elements could also be taken up out of the air was first settled affirmatively in connection with Spanish moss (*Tillandsia usneoides*). Today it is the general custom in California and Florida to bring zinc and other trace elements to the plant, not in the fertilizer through the root, but by putting it on the leaves, since the leaves take up the fine substances very well, better indeed than do the roots.

It was discovered that one-sided mineral fertilizing impoverishes soils and plants with regard to trace elements. Perhaps the most important discovery was that adding trace elements to the fertilizer does not mean that the plants can always take them up. The presence or absence of zinc, in a ratio of one part to one hundred million determines whether an orange tree produces healthy fruit or not. In the years from 1924 to 1930 people ridiculed the biodynamic preparations "because one certainly cannot influence the plant with high dilutions."

We referred to zinc in this connection because on the one hand this trace element is so exceptionally essential for the health of a number of plants, and also for their yield — precisely in the highest dilutions — and on the other hand it is especially accumulated in fungus growths. A remark of Rudolf Steiner's points to an interesting interrelationship that can really only be properly understood in the light of research of the decade since the end of the Second World War:

> The harmful parasites go hand-in-hand with the
> fungi ... in this way plant diseases arise and also
> coarser abnormalities ... We should actively encourage
> the mushrooms and toadstools to grow in these [wet]
> meadows. You will then experience the remarkable
> fact that if you have even a small area where
> mushrooms are growing, their relationship to the
> bacteria and other parasitic creatures will keep these
> creatures away from everything else ... In addition
> to what I already described for combatting such
> plant pests, it is also quite possible to keep harmful
> microorganisms away from the farm simply b y
> establishing some wet meadows. (*Agriculture*, p. 147).

Among the fungi and fungus-like organisms belong also the so-called fungi imperfecti and a botanically intermediate member, the ray fungus or Actinomycetes and Streptomyces from which antibiotic remedies are extracted. I found that these organisms have a quite special share in humus formation and the rotting processes, and are accumulated to a high degree in the biodynamic fertilizer preparations. These preparations also accumulate a number of the most important trace elements, such as molybdenum, cobalt, zinc and others whose value is today recognized on the basis of experiment.

Regarding soils a strange situation was discovered. Analysis of available plant nutrients showed that these figures varied in one and the same soil at different times of the year. Seasonal and even daily variations were found. These variations are often larger within the same test area than the differences between neighboring fields, a good and a poor soil. But seasonal and daily variations are the result of the position of the earth within the planetary system, they are of cosmic origin. In fact times of the day and times of the year can variously influence the solubility and the availability of the nutritive substances. In the physiology of plants and that of animals (glandular secretions, hormones) numerous phenomena are subject to such influences. The leaf of the Bryophyllum contains oxalic acid whose concentration follows the time of day almost like the hand of a clock. Although here and in many other cases the substantial nutritive basis is the same — plants of the same species can, under the influence of different light rhythms and cycles, have an entirely different intake and loss of substances. Joachim Schultz, a research scientist at the Goetheanum who regrettably died relatively young, had begun to test experimentally an important statement of Rudolf Steiner's, namely that the effects of the light in the morning and evening hours influence plant growth differently (beneficially) from the light of the noon and midnight hours (hinderingly).

Observing Schultz's experiments, I was struck by the fact that plants grown on the same nutritive solution showed completely different synthesis of substance, nitrogen for example, depending on the light rhythms to which they were exposed. Those exposed in the morning and evening hours showed a rich growth, benefited by nitrogen effects. Those under midday light only, were stunted and showed deficiency symptoms. With these tests the way was now experimentally prepared for demonstrating that the "cosmic" effect of the light and warmth of the sun especially but also of other sources of light, controls events in the field of substance. They regulate the course of changes of substance. When and in what direction these take place, and to what extent the total growth, the form of the plant is influenced thereby, depends on the cosmic constellation and source of energy. The latest research in the field of photosynthesis too is of a nature that may open the eyes of materialistic circles to such occurrences. It would seem to us that in this respect, too, Rudolf Steiner was a forerunner who prepared the way for a new direction in research. It would take up too much space to tell of all the phenomena already known. But it is no longer possible to simply dismiss the influence of the cosmic forces with the word "superstition." It is impossible to do that as soon as one takes into consideration the physiological and biochemical dependence on cosmic forces of the metabolic function, the soil life, the sap flow, and especially also the processes taing place in the plant root.

There was an older way of looking at nature, based in part on mystery tradition, in part on instinctive clairvoyance, extending from the time of Aristotle and his pupil Theophrastus the botanist to the period of Albertus Magnus and the late medieval teaching of "signatures." In that approach to nature, relationships were spoken of between the individual plant species and specific cosmic constellations. These constellations are the creative impulses under whose influence the species differentiated themselves, and different forms of being arose. If one considers that cosmic rhythms have such a significant influence on the physiology of metabolism, on the function of the glands, lymph stream and turgor, then it is only a small step to the next perception — the conscious and scientific experimental study of the creative effects of the constellations. Numerous collaborators of Rudolf Steiner have already demonstrated the effect of the formative forces through experiments (capillary dynamolisis method by Lily Kolisko), or in plant and crystallization tests (Hans Krüger, Frieda Bessenich, Alla Selawry, among others).

Regenerating seeds

A special task developed out of Rudolf Steiner's suggestions for plant breeding. The investigations in this connection were carried out by myself and other collaborators (Immanuel Voegele, Erika Riese, Martha Künzel, Martin Schmidt) partly in collaboration, partly independently. Starting from the basic idea of the cosmic, creative constellations, in every species or subspecies the original impulse slowly ebbs away and becomes lost. This original impulse is inherited in the plant as formative force, being transmitted by certain organs (chromosomes for example). One-sided fertilizer material gradually drives out the after-effects of the original forces, so that the plant becomes "weaker." The seed quality degenerates. This was the problem that was first brought to Rudolf Steiner and which stimulated the creation of the biodynamic method.

The task was to give the plant, as a system of forces under the influence of cosmic effects, back again to the whole of nature. Rudolf Steiner indicated that a number of the "worn out" plants, those that had become alienated from their origin, are degenerating at such a rate that by the end of the twentieth century one will no longer be able to rely on them for field cultivation. Wheat and potatoes, among others, were mentioned, but other crops such as oats, barley, and alfalfa, too, come into this category. Ways were outlined by which, from wild, still unexhausted relatives of the cultivated plants, new varieties with viable seeds could be grown. A start has been made on these tasks with good results. Today [1956] there are new varieties of wheat in existence. Martin Schmidt did significant work to show the rhythm of seed arrangement in the grain ear, and especially to demonstrate the difference between a food plant and a seed plant. According to Rudolf Steiner this basic difference in characteristics depends on whether the seed is sown nearer the winter or the summer season. In the development of the proteins, amino acids, phosphorus lipoids, enzyme systems, etc., the biochemist will one day also be able to follow these differentiations of substance, through chromatography.

The degeneration of wheat has today become a fact. Even on good soils, the protein content is reducing (from 13% to 8% in the case of red wheat in a number of regions of the United States during the last thirty years). Whoever grows potatoes knows how difficult it is still to raise a healthy potato not attacked by insects and viruses — not to mention the question of flavor. Biodynamically raised wheat has maintained itself on the high protein level. The promising work in potato breeding was unfortunately interrupted by war and other disturbances.

From the dynamic point of view, the problem of pests is one of the most interesting and instructive. A biological balance is disturbed, and as a consequence there is a degeneration. There is an onset of pests and diseases. Nature herself liquidates that which no longer has the capacity for life. So pests are actually a warning from nature that the original forces are being lost and that one has transgressed against a state of balance. For this warning, American agriculture today [1956] is paying, according to official figures, five billion dollars annually in crop losses, plus an additional seven hundred and fifty million for pesticides. That the goals are not achieved with insect poisons, and that, despite the poisoning, new and resistant pests are emerging, is something that is beginning to enter people's consciousness. For example, the most advanced scientists, like William Albrecht of Missouri, has shown that by one-sided fertilizing the balance of protein and carbohydrate in the plant cell has been disturbed to the detriment of the protein and protective layers on the outside of the leaves. As a consequence the plants become "tastier" to the pests. Gaining the insight that insect poisons only preserve what is beginning to become a "corpse of nature" but cannot hold back the general dying off, is a bitter experience. Already leading entomologists are making their voices heard, opposing chemical attempts to control pests and the disturbances of health connected with this approach, and are advising biological controls. Yet biological control, according to the instructions given, for example, by American research stations, is only then possible when one applies no poisons and tries to re-establish the natural balance. That health and resistance to diseases and pests is a function of the biological balance, including the cosmic factors, had already been explained by Rudolf Steiner in the Agricultural Course. Here too it is plain to what an extent this spiritual-scientific way of thought, based on Goethe, was ahead of its time.

The Biodynamic Preparations

The biodynamic method is a system of measures to be taken in farming and gardening with the goal of improving the organic and humus state of the soil.

The conservation of organic matter is one of its main concerns. Where there is no organic matter available there is no possibility of introducing biodynamic practices.

Compost and manures are the main sources of organic matter which can be added to soils in order to improve them. Aside from these there are leaf mold, wastes, etc.

It is the first endeavor of the biodynamic farmer or gardener to collect as much available raw material for compost and to produce as much manure as possible. Biodynamic farming emphasizes the full economic utilization of increased livestock.

Manures and compost are not left to a chance fermentation, since losses may occur of valuable minerals, organic matter, or nitrogen. Scientists and those who practice biodynamics have carefully studied the conditions under which manure and compost rot to produce humus rich in nitrogen and preserving its total original content of organic matter and minerals. A routine procedure for building compost and manure heaps has been worked out and described.

Experiments have shown that the fermentation of manures and compost is effected by bacterial action. Losses of nitrogen are brought about by ammonifyers, denitrifyers, and heating (thermophile) bacteria. Undue oxidation may take place through too much access of air, or there may be washing out by exposure to rain. An acid peat or muck-like organic matter may result under anaerobic, too moist conditions.

The conditions are described under which the proper rotting or fermentation takes place, and procedures in farming and gardening are introduced in order to reproduce these favorable conditions. This is one of the major points stressed by the biodynamic method: how to bring about a proper fermentation of compost and manure without losses.

It has been observed that bacterial action as well as fermentation processes can be influenced by growth hormones, enzymes, traces of certain types of natural humus, extracts of certain plants, etc. In order to influence the humus-forming process and "digestion" of raw materials the biodynamic preparations have been introduced. These contain particular bacteria, typical for instance of a fertile humus soil, which are also found in earthworm castings, and growth-stimulating substances (auxins, growth hormones). The method of producing these preparations was described by Rudolf Steiner.

Originally the formulas of the preparations were kept secret, not because a "secret society" stood behind them, but because it was desired to avoid misuse and misinterpretation until there was sufficient scientific proof of their action. Later more details about the composition of these preparations were made known.* However, the raw materials used for the preparations have always been stated openly. The materials undergo fermentation in the earth, some of them wrapped in animal organs in order to concentrate bacteria, growth substances and enzymes. There are nine different preparations that have been numbered 500 through 508.

Preparation 500 (horn manure)

Preparation No 500 is made from fresh cow manure packed into cow horns and left for six months' fermentation in the earth.

---

* Pfeiffer wrote this article in the late 1940s, the time when details of the preparations became more widely known.

Bacteriologically, fecal bacteria disappears during this period and a microflora accumulates which is very much the same as that of earthworm castings, That is, humus forming bacteria. Chemically, we find an increase of nitrate nitrogen from 0.06% to 1.7%, or about 28 times the original content.

The preparation is diluted in water and stirred for one hour; during which an increase of up to 75% oxygen absorption can be found. It is sprayed directly and immediately on the land. The microflora which it contains is thus distributed evenly over the land. The important effect is a stimulation of root growth, particularly of the fine hair roots, and of humus forming processes in the soil. Since the microflora is similar to that of earthworm castings, preparation No 500 acts as a kind of substitute earthworm. The bacteria count is 500 million aerobic bacteria per gram.

Spectrographic analysis of No 500 shows the following results:

alum up to 10%
boron 0.01%
barium between 0.01 and 0.1%
calcium from 1.0 to 10%
chromium 0.001 to 0.01%
copper 0.001 to 0.01%
iron 0.01 to 1.0%
magnesium 1%
manganese 0.001 to 0.01%
molybdenum 0.001%
sodium 0.1 to 1.0%
phosphors between 0.01 and 0.1%
lead 0.001%
silicon between 1.0 and 10%
titanium 0.01 to 1.0%
vanadium 0.001%.

It is evident that No 500 made from manure produced by various feeding methods and pastures will vary somewhat. We prefer, therefore, to use the cow dung for it when the cows are feeding on the very best pasture with lots of clover and alfalfa in it. Also there might be a variation in quality as to soils on which the feed was grown and to seasonal changes. We use only the dung from cows which feed on previously biodynamically treated fields and their crops. The preparation is started in fall, September to October, when it appears that the pasture feeding is most concentrated, and the animals also eat leaves from shrubs in the bordering hedgerows. The manure at this time is rather firm and contains more minerals than in spring and early summer when it is more juicy and green.

Preparation No 501 (horn silica)

Preparation No 501 is made from rock crystal, finely powdered, treated for 6 months together with cow horns in the earth.

The research on the bacteriological processes is not quite concluded. The growth stimulating effect of this preparation upon stem and leaf and the assimilation process in plants has been observed. For example, it increases the assimilation in sunflower leaves up to 3.5 times (measured in weight increase). The bacteria count of the quartz powder, none; of the preparation No 501, 40 million aerobic.

Chemically, the preparation of the quartz powder, compared with the untreated quartz powder, shows an increase of nitrate nitrogen from zero to 0.007%. Also magnesium, potassium (0.01%), phosphates (0.01%) show up. The spectrographic analysis reveals quite a few changes. We observe silver, 0.001% before and after the treatment, as well as copper, 0.001%. Alum is increased from 0.001% to a little more than 0.1%. Boron remains unchanged at about 0.01%. Calcium is considerably increased from a faint trace to as much as 1.0 to 10%. A trace of chromium appears after preparation (0.001%). Iron is increased about five times to 1%. Magnesium is increased about 100 times to 0.1%, manganese from a faint trace to 0.01%. A trace of molybdenum appears, sodium in amounts from 0.1 to 1.0%, phosphorus from 0.01 to 0.1%, lead up to 0.01%. Traces of titanium, vanadium and zirconium are also found. The main element of No 501 is silicon, — silicium dioxide being the major compound of the preparation, — comprising about 90%.

Preparation No 502 (yarrow)

This preparation is made from yarrow blossoms (*Achillea millefolium*), fermented together with deer bladders over a period of six months in earth during the winter. The analysis of the available minerals shows following:

> a decrease of potassium from 1.05% to 0.13%
> an increase of calcium from 0.05% to 0.375%, or 75 times
> a slight decrease of magnesium from 0.01% to 0.005%
> phosphate remains stable at about 0.06%.

The major increase is again observed in nitrate nitrogen from 0.07% to 2.5% or about 36 times the original. Nitrogen fixing bacteria have migrated into the preparation and have lived and worked there. According to Rudolf Steiner, No 502 has a stimulating effect on the use of sulphur and potassium by plants in their growth. This in turn effects the building up of protein and carbohydrates and their balance. This preparation, as well as the others, acts as a biocatalyst.

Spectrographic analysis of this preparation, compared with untreated yarrow blossoms, reveals the following contents, in addition to those already described:

> silver, a faint trace, remains unchanged
> alum at 0.1% has slightly decreased through the fermentation
> boron with 0.1%, chromium .001%, manganese 0.01%, lead 0.001%, silicon 1.0%, remain unchanged
> iron increases from 0.1 to 1%.

The spectrographic analysis shows a higher percentage for

available minerals than the chemical analysis but the rate of decrease is the same. Molybdenum shows up with 0.0001% after the preparation, so does nickel at 0.001%, vanadium 0.0001% and zirconium with a faint trace. Titanium is enriched to 0.1%, about 10 times.

The bacteria count, as determined by the plate counting method on beef-agar peptone, after 48 hours at 29°C, of the raw material (yarrow blossoms) is 30 thousand aerobic, no anaerobic bacteria per gram, that of the final preparation is 910 million aerobic bacteria, anaerobic none. The microflora of the yarrow consists mainly of sarcina and micrococcus and similar dust bacteria, while the final preparation has an entirely different flora, mainly of actinomycetes and bacteria belonging to the Bacillus type.

Preparation No 503 (chamomile)

This preparation is made from *Matricaria chamomilla*, chamomile blossoms, fermented together with the small intestines of healthy animals (cows), again fermented over the winter bedded in earth. It is clear that only good humus earth can be used, for its richness in soil-building bacteria which migrate and accumulate in the preparations. No "dead" earth should be used. The animal organs' content of hormones and growth substances is most important, therefore only healthy animals should be used. When we say healthy, we mean not only animals which appear to be healthy but those which are free from malnutrition, deficiency diseases, infections, and which are not one-sidedly forced to too high production. Their feed too should be derived from fertile, healthy soils. The chemical test for available minerals has revealed that in this preparation a considerable increase in calcium, nitrate nitrogen and phosphate took place during the fermentation:

calcium from 0.05% to 0.41%, almost 10 times
nitrate nitrogen from 0.04% to 3.1%, 77.5 times
phosphate from 0.08% to 0.75%, about 9.4 times

potassium has declined two thirds
magnesium declined 2.5 times
ammonium nitrogen is very low, 0.002%, and remains
    constant.

By "available" minerals we mean extractable with a buffered, slightly acid organic acid (acetic and citric acids). Spectrographic analysis reveals all the elements present, no matter whether they are available or not. Its findings are:

constant before and after treatment:
silver faint trace
boron 0.1%
barium 0.1%
chromium 0.001%
copper 0.01%
manganese 0.01%
increases:
alum about five times to 1–10%
calcium 10 times
iron to 1.0%, 10 times
molybdenum from zero to 0.001%
sodium about five times to 1.0%
nickel from zero to 0.001%
lead from zero to 0.001%
silicon 10 times to 1.0%
titanium from 0.001 to 0.1%, or about 100 times
vanadium from zero to 0.001%
zirconium from zero to 0.001%.

The total bacteria count (plate method on beef-agar peptone, after 48 hours at 29°C) before fermentation of the chamomile blossoms was 90 million bacteria per gram; after the treatment it was 800 million per gram, all aerobic. There was no anaerobic growth.

The preparation No 503 is of special interest because it is made from chamomile. This plant contains a growth hormone which is a particular stimulant for the growth of yeast. The remarkable thing about this growth hormone is that it works in very high dilutions. F. Boas of the University of Munich reported, before the Second World War, that chamomile juice is active at its best in a dilution of 1 to 8 million. Since then, it has been reported that even in dilutions of 1 to 1 billion these hormones are still effective. That high dilutions could affect plant growth seemed unbelievable in the 1920s and 1930s when the biodynamic preparations were first tried. The study of growth substances and trace elements has revealed in the meantime that dilutions of 1 to 1 million up to 1 to 1 billion are nothing extraordinary. In fact, many of these biocatalysts do their jobs best just in these high dilutions while in a more concentrated form they may be ineffective or even harmful. This is the case with trace elements such as boron, cobalt, molybdenum and many others. We will report about related experiments in detail below.

Preparation No 504 (nettle)

This preparation is made from stinging nettle (*Urtica dioica*) leaves and stems, buried about two feet (60 cm) deep in humus earth, separated from the soil by a thin layer of peat moss. They undergo a fermentation during one year which produces a fine black-brown humus and breaks down the fibrous parts of the plant. The plants are used shortly before the blossoming stage, no mature plants should be used. Steiner claimed that this stinging nettle compost if inserted in compost or manure heaps counteracts undue fermentation and break-down processes and protects against losses of nitrogen. The stinging nettle is an interesting herb, rich in vitamins and iron. In old times it was used as a remedy against anemia and in order to strengthen the vitality. The stinging hairs contain formic acid. The interesting change during the preparation is that the stinging nettle humus is enriched about 100 times in molybdenum and vanadium which are the trace elements necessary for the activity of nitrogen fixing bacteria.

Chemical analysis of available elements reveals a washing out of potassium from 1.2 to 0.1%.

Spectrographic analysis shows the maintenance of:

   silver, faint trace
   boron 0.1%
   chromium 0.001 to 0.01%
   copper 0.01%
   manganese 0.001 to 0.01%
   calcium between 1 to 10%
   magnesium 5%
   lead between 0.001 and 0.01%

and shows following increases:

   alum increases considerably, 10 times to 1.0%
   sodium 10 times to 1.0%
   nickel from zero to 0.001%
   titanium from 0.01% to 0.1%
   zirconium from zero to 0.001%
   iron from 0.1 to 1.0%

Barium declines slightly to 0.001 from 0.01%. In spite of the stimulation of nitrogen fixation which goes out from this preparation, its own nitrate content, of the fermented product, is half that of the green leaves (about 0.73%). The available phosphorus maintains its level.

The bacteria count of the stinging nettle was 60 million aerobic bacteria per gram; anaerobic TMC (too many to count). The final preparation count was 1,050 million aerobic, 470 million anaerobic.

Preparation No 505 (oak bark)

This preparation is made from oak bark (*Quercus robur* or *Quercus alba*) from not too old a tree, buried together with the skulls of ruminants. The resulting humus is dark blackish brown and of a particularly fine structure. The peculiarity of the oak tree is that it accumulates a tremendous amount of calcium in its bark during growth even when it grows in a soil very poor in calcium. In fact, the highest calcium ash content was found in oak trees growing on a sandy, calcium deficient soil. The calcium content of this preparation is therefore very high, more than 10%, as determined spectrographically. Of these 10% about 0.1% is readily available. Potassium decreases slightly from 0.05 to 0.013%; phosphate increases from 0.01 to 0.03% of the available fraction, while the spectrograph shows an increase of from 10 to 100 times.

The main increases in this preparation are
   alum from 0.1 to above 1%
   iron from 0.1 to 1%
   magnesium from 0.1 to 1%
   molybdenum from zero to 0.0001%
   sodium from a faint trace to between 0.1 to 1%
   nickel from zero to 0.0001%
   silicon from 0.1 to 1%
   titanium from 0.01 to 0.1% (10 times)
   vanadium from zero to 0.0001%
   zinc from zero 0.001%
   sulphates from zero to 0.005%.
The following elements or compounds remain unchanged:
   iron 0.01%
   manganese 0.01%
   lead 0.001%
   nitrate nitrogen 0.073%
   chromium 0.01%.

The bacteria count of the powdered bark was: aerobic 30 million, anaerobic none; of the final preparation aerobic 2 billion, anaerobic 70 million.

This preparation in its effects stimulates the resistance of plants to disease.

Preparation No 506 (dandelion)

This preparation is made from dandelion (*Taraxacum officinalis*) combined with the mesentery of ruminants, rich in glandular cells. There is a decrease of potassium from 1.25 to 0.27%. The increases through the fermentation process are quite remarkable. Of the available fractions we have an increase of calcium from a trace to 0.08% while the total calcium is around 10%; an increase in available magnesium from 0.007 to 0.01%, the total being 10%; increase of nitrate nitrogen from 0.336 to 3.38% through bacteria action; increase of phosphate from 0.46 to 0.72%. Sulphates show up with 0.005%.

Spectrographic analysis runs as follows:

increase of alum from 0.1 to 1%
iron from 0.1 to 1%
molybdenum from zero to 0.0001%
nickel from zero to 0.0001%
titanium from 0.1 to 1%
vanadium from zero to 0.0001%
zinc from zero to faint trace.

The following remain unchanged:

boron 0.1%
barium 0.1%
copper 0.01%
manganese 0.01%
sodium 1%
lead 0.001%
silica 1%.

There were slight decreases in chromium, 0.001% to faint trace, in silver from faint trace to zero. In general, it can be stated that the decreases in all of the above preparations are due to washing out, while the increases are the result of migration and absorption from outside and of bacterial action.

The bacteria count of the dandelion was 70 million aerobic, no anaerobic; of No 506, 360 million aerobic plus a spreading species, and 180 million anaerobic bacteria per gram.

Preparation No 507 (valerian)

This preparation is the pressed out extract from valerian blossoms. *Valeriana officinalis* is commonly used as a remedy for spastic nervous conditions. The chemical findings of the juice were:

> potassium 0.335%
> calcium 0.425%
> magnesium 0.005%
> nitrate nitrogen 0.145%
> ammonia nitrogen showed an increase from 0.02% to 0.6%
> phosphate 0.062%
> manganese slight trace.

The valerian preparation (507) has been used in greenhouses where it stimulated assimilation and increased (that is, darkened somewhat) the colors of flowers. It is a valuable help in greenhouses in order to compensate for the reduced effect of the light which passes through the glass.

No 507 is used in high dilution (5–10 drops per 2 gallons, 7.5 litres, of water) as a spray over the base and cover of manure and compost heaps. Previous experiments (see E Pfeiffer, *Soil Fertility*, Chapter 11) have shown that it attracts earthworms and stimulates their propagation. It is the cold-pressed juice of *Valeriana officinalis* blossoms.

Spectrographic analysis reveals the following contents:

> aluminum 0.001%
> boron 0.001%
> barium 0.001%
> calcium 1.0 to 10%
> copper 0.001%
> iron 0.001%
> potassium 0.01%

magnesium 0.1 to 1.0%
manganese 0.01%
sodium 0.001%
phosphorus 0.1%
silicon 0.001%

The bacteria count was 1 million aerobic, anaerobic none.

Preparation No 508 (horsetail)

This is the dried herb *Equisetum arvense*, horsetail. Its mineral contents are:

silicon above 10%
potassium 1.15%
calcium 0.42%
magnesium 0.01%
nitrate nitrogen 0.47%
anmonia nitrogen a trace
phosphate 0.06%
manganese a small amount
sulphates 0.36%.

Spectrographic analysis of *Equisetum arvense*, besides the above mentioned "availables," shows:

silver faint trace
aluminum 0.1%
boron 0.001%
barium 0.01 to 0.1%
chromium trace; iron 0.1 to 1.0%
potassium 1.0%
magnesium 1.0 to 10%
manganese 0.01%
sodium 0.1 to 1%
lead faint trace
silicon above 10% (it is the major element in this plant)
titanium trace.

Calcium is usually very low in equisetum but in the plants we use it was exceptionally high, being second to the silicon content.

The bacteria count was 21000 aerobic bacteria per gram, no anaerobic bacteria.

The place where all these plants are grown, again has an influence upon their contents. One should be sure to use only those growing on the very best, rich, if possible, biodynamically enlivened soils.

The average concentration of all of these preparations used as a starter or stimulant is 0.005%. This may seem low, but if one considers the hormone effects as described in current literature, being effective in concentrations of 1 to 100 million and more, then these are still "concentrated."

Effect of the preparations on yeast growth

The following test was carried out by a commercial yeast fermentation research laboratory through the courtesy of a friend of the Biodynamic Farming & Gardening Association. The preparations were sent, identified only by numbers, in dilutions of 1:1000 to the laboratory. The report of the laboratory follows:

Four samples labeled I, II, III and IV were tested for their effect on yeast growth by adding various amounts to a synthetic basal medium. The synthetic medium had the following composition:

Moyer salts*	3.5g	$MgSO_4$ KH_2PO_4	0.25g 0.50g
Dextrose	50.0g	$NaHO_3$ Zn	3.00g 0.10g
50ml of a solution containing .01% each of beta alanine, calcium pantothenate, inositol and thiamine chloride			
Water to give 1000ml			

* These salts are mixed in the proportions shown and were originally compounded for mold work. Zn probably not required for yeast.

This medium tubed in 10 ml amounts and sterilized for 20 minutes at 15 psi (103 kPa) steam pressure. The following dilutions of the yeast stimulants (which were sterilized 5 minutes at 10 psi (69 kPa) before using) were made: 1 to 10,000, 1 to 100,000 and 1 to 1,000,000. The tubes after having the stimulants added to them in the quantity indicated were inoculated with .05 ml of a 24 hour culture of yeast and incubated for 24 hours. Turbidity measurement of the resultant growth was made and these figures were referred to a curve previously plotted against yeast cell counts. The following results were obtained on sterilized stimulants:

Yeast Stimulant	1:10,000	1:100,000	1:1,000,000	No Stimulant Control
I	4,166,000	3,834,000	3,166,000	
II	3,332,000	3,664,000	3,498,000	2,000,000
III	3,900,000	4,166,000	3,498,000	
IV	3,834,000	3,900,000	3,166,000	

This is the key to the numbers: **I** was a 1% boiled extract of the chamomile blossoms. The concentrations applied correspond, therefore, to 1 to 10,000 of 1%; 1 to 100,000 of 1%; 1 to 1,000,000 of 1%. The dilutions of numbers **II, III** and **IV** applied were made from a dilution of 1to 1,000, the final dilutions therefore being 1 to 10,000,000; 1 to 100,000,000 and 1 to 1,000,000,000. **II** was preparation No 503 (made from chamomile); **III** was a mixture of all the preparations from 502 to 506; **IV** was preparation 507.

The remarkable result of this test is first, that yeast growth was considerably stimulated even with sterilized preparations. This points to the presence of heat resistant hormones. And second, the effect is lasting, no matter whether a dilution of one to a million or one to a billion is applied. In the case of the mixture of all the preparations the highest figure — more than double the control —

was obtained with a dilution of 1 to 100 million, while in the case of the extracted chamomile a slight decline was observed in the higher concentrations.

In practice the preparations are used at 2 grams each of 502 through 507 to 15 tons of compost or manure, in a concentration of 1 to 7.5 million, certainly not too high a dilution in view of the above results. No 500 and No 501 are sprayed directly on the land and on the plants, resp. Figuring the top layer of the soil of an acre to weigh about 2 million pounds, and using 40 grams of No 500 per acre, the concentration would be 1 to 25 million, which is almost exactly the concentration of the optimum yeast fermentation test. In the field of growth hormones, such dilutions or concentrations are not at all unusual. Of course, if we think in terms of fertilizer as was the custom at the time of the invention of these preparations in the 1920s one would marvel at such high dilutions and say, as was said then: How can such a speck of material have any effect at all? The subsequent progress of science itself has taught us better.

Now why do we emphasize the trace element contents of these preparations? First of all, one must remember that the process of fermentation increases the content of some of them, such as molybdenum, manganese, titanium, and vanadium. Recent research has shown that traces of these elements in a concentration of 1 part per million are essential to the proper growth and nutritive value of plants and bacteria.

Molybdenum and vanadium are necessary for the action of nitrogen fixing bacteria like *Azotobacter chroococcum*. Without these traces the bacteria would not do their job, which is absorption and fixation of atmospheric nitrogen and its transformation into nitrate nitrogen. That they do work in the presence of these trace elements we can see in the remarkable increases of nitrate nitrogen in No 500, 28 times; No 502, 36 times; No 503, 77.5 times; No 506, 10 times.

Manganese has an influence upon the reduction-oxidation

process in plant leaves. The roles of other trace elements still need more study. A study still in progress of the bacterial changes in and through the preparations shows that through the treatment, fecal bacteria in manure and those which break nitrogen down to ammonia or nitrates with subsequent losses in nitrogen, disappear and give way to nitrogen fixing bacteria.

Fermentation studies

Fermentation studies were made under sterile conditions, allowing only the bacteria which were present in fresh cow manure, the additions of preparations No 500, No 502 to No 507, and earthworm castings to work.

The following mixtures were prepared and filled, under sterile conditions, in Blake bottles of 1000 ml capacity.

1. & 2. Fresh cow manure, 200 g plus 40 ml sterile distilled water, untreated.
3. Fresh cow manure 200 g plus 0.2 g of No 500 diluted in 40 ml sterile distilled water.
4. Fresh cow manure 200 g plus 40 ml sterile distilled water, 20 g air dry earthworm castings.
5. Fresh cow manure 200 g plus the same as No 4 plus 0.2 g No 500.
6a. Earthworm castings alone 200 g plus 40 ml sterile distilled water.
6b. Earthworm castings 200 g plus 0.2 g No 500, 40 ml sterile distilled water.
7. Cow manure 200 g, 40 ml sterile distilled water, 0.2 g. of each of No 502 to No 507, 20 g earthworm castings.

These materials were kept under sterile conditions. One set of each of the bottles was set upright, the other horizontally with the narrow side down. Fermentation started at once. The surface layer turned into a brown humus after ten days to six weeks. Under the conditions of the experiment this layer dried up somewhat and sealed up the lower layers. In this top layer we had a typical aerobic fermentation, while in the sublayer a largely anaerobic fermentation started, chiefly sponsored by gas-forming Clostridia and other bacteria. The bottles were purposely kept quiet, not shaken, stored in the dark. The top layer was relatively thin, only a few millimeters, while the bottom layer was several inches thick in the upright bottles, one to two inches in the horizontal ones. This sublayer remained greenish in color throughout the experiment.

Besides the bacteriological changes it was observed that a grey mold developed in all the untreated bottles while this was inhibited on the biodynamically treated materials. A considerable shifting of chemicals took place.

Potassium: accumulated in the top layer of the control and with the No 500 treatment. Only with the No 502–507 treatment did it accumulate in the bottom layer. This is in accordance with the findings of the treatment of experimental heaps in the open air.

Calcium: With No 500 and No 502–507 treatment it migrates slightly into the sublayer; with earthworm castings, treated and untreated, into the top layer. There was also a slight increase of calcium, while in the untreated manure a slight decrease was observed.

Magnesium: showed slight rather unspecific variations.

Nitrate nitrogen: showed considerable increases in almost all cases, even the untreated manure showed a strong increase. These increases rose steadily during the first 6 weeks to 3 months, then a decline set in so that after 10 months much was lost again. Here a peculiar phenomenon was observed. In the case of untreated manure and earthworm castings or mixtures of both the better increase was in the sublayer, probably due to the action of *Clostridium pasteurianum*, while the biodynamically treated manure showed the better rate of increase in the top layer. This result is confirmed by the open air experimental heaps, where the highest increase was also in the top (aerobic) layer with the treated heaps, while in the anaerobic bottom layer in the open air heaps, under unsterile conditions considerable losses were observed. The increase in the upper 5 inches (12 cm) of the open air biodynamic heap treated with 502–507 was about three times the original nitrate content.

Ammonia nitrogen: Only traces were observed in all cases. This differs from the open air experimental heaps where a considerable amount of ammonia appeared in the sublayer of the treated (502–507) heaps and in the top layer of an untreated heap set up on a stone base.

Phosphates: showed a small rate of fluctuation. The No 500 treated bottled manure and the earthworm castings showed a small accumulation in the top layer, while the 502–507 treated bottled manure showed a high rate of accumulation in the top layer. In the open air experimental heaps with 502–507 a slightly better accumulation was observed in the top layer, 14 lb/ton (7 kg/t) as against 10 lb/ton (5 kg/t) in the bottom layer. In the untreated heap on the stone base the accumulation was in the bottom layer (20 lb/ton, 10 kg/t) as against 12 lb/ton (6 kg/t) in the top layer.

All analyses were repeated several times since differences in sampling may occur. If two tests gave similar results a mean of two or three analyses is reported. If higher fluctuations were observed, then 5 to 6 tests were made to get a better average. The reliability of the tests was within the limits of about 2%.

The bacteriological findings are rather complicated, the isolation and identification of the bacteria require much time. The following table gives the changes calculated in lb per ton (in the laboratory we usually test for parts per million). Aerobic in the table is comparable to the top layer of a heap, anaerobic to the bottom layer.

Element lb/ton*	Fresh cow manure	Cow manure untreated after 3 mnths		Cow manure plus No 500 after 3 mnths		Cow manure plus earthworm castings		Cow manure plus earthworm castings plus No 500		Earthworm castings		Cow Manure plus earthworm castings plus No 502–507 10-11 mnths	
		aerobic	anaerobic	aerobic	anaerobic	aerobic	anaerobic	aerobic	anaerobic	a. alone	b. No 500	aerobic	anaerobic
pH	7.0	7.8	7.8	8.0	8.0	7.8	7.8	7.8	7.8	6.5	6.5	8.0	6.8
Potassium	8.4	19.8	5.28	23.10	19.8	8.58	9.24	13.2	11.0	0.26	0.24	9.2	32.0
Calcium	7.5	6.8	6.8	7.2	8.3	9.13	8.36	9.9	8.36	1.5	1.5	6.0	7.6
Magnesium	0.22	0.22	0.22	0.198	0.198	0.27	0.22	0.22	0.20	0.176	0.22	0.25	0.17
Nitrate nitrogen	0.2	3.05	5.94	5.83	7.01	1.48	2.46	3.41	0.88	0.67	1.85	80g	36g
Ammonia nitrogen	trace	0.02	trace	trace	trace	trace	trace	slight tr.	slight tr.	0	0.03	trace	13g
Phosphates	18.0	15.0	13.75	17.6	14.3	15.04	17.05	18.7	16.5	6.15	5.50	20.24	4.4

* 1 lb per US (short) ton = 0.500 kg per (metric) tonne

This experiment demonstrates that the preparation No 500 and to a certain extent earthworm castings also, increase the amount of available minerals. The formation of the nitrates, however, is a fluctuating process. Even in untreated manure we have an increase of nitrates as long as the manure is covered and protected against burning or drying out. The increases with the addition of the biodynamic preparations in the bottle series were somewhat better than in the untreated series. However, if stored for a longer period without contact with the earth (or cover) the nitrates might be broken down again and lost as is shown in the case of No 7, where the test was made after ten months.

On the other hand we observe a tremendous increase in nitrates during the first six weeks to three months of fermentation and, if the heaps are covered with earth, these nitrates will stay. It lies entirely in the hand of the compost builder whether his heaps will gain, maintain, or lose their nitrate level. Here the open air experimental heaps fare better than the bottle series under sterile conditions.

The fermentation series reported upon so far, were only made in order to obtain a first basic orientation to what happens under completely controlled laboratory conditions. Therefore, we excluded other additions to the fermenting manure which are usually present, such as earth, litter, etc. We wanted to see what the manure alone would do, with and without the starter stimulants. The organic matter of such additions is of course an important energy source for the microlife, while in the closed bottles this material is consumed after some time and no more growth is possible.

It must be realized that pure manure contains a lot of water; if fresh, green, and juicy, up to 90% or more. About 25 to 30% of its solid matter consists of bacteria bodies. This is a very living and changeable substance. No wonder that we find there a continual fluctuation of more or less available minerals, according to the life process which is prevalent at any given period. Under certain conditions only, can a stable form or balance be reached.

Conclusions

From these preliminary studies and others not reported here with experimental heaps in the open air and in contact with the earth, the following conclusions are drawn. A favorable fermentation with the accumulation of potassium and phosphate and the fixation of nitrate nitrogen takes place in the aerated layer of manure under the treatment with No 500, and No 502 to No 507. In the not aerated or even entirely anaerobic layers unfavorable tendencies become more prevalent in the course of time. The nitrate nitrogen increases for about three months after the heap is set up, nearing a peak around six weeks. The nitrates remain if the heap is covered with a thin layer of earth. In the sublayer unstable conditions exist.

In practice, the result is that a piling up of not too large manure heaps is preferable so that air can have access to all layers of the heaps. The covering with or the insertion of thin layers of earth is essential. In the case of large heaps or very wet manure, special air channels or base drainage must be provided. This, incidentally, is the same advice which we gave long ago in practice, on an empiric basis.

The mix of manure

Nitrogen fixing bacteria need carbohydrates as sources of energy. If, in a closed system (such as bottles No 6 and No 7), the carbohydrate sources are consumed then there will be no further development. Manure should be mixed with litter. The question is what amount of litter, straw, etc., gives the most favorable results. We cannot as yet tell exactly which are the best proportions in spite of a great deal of talk about the nitrogen to carbon relationship.

Empirically, we chose the following mixture:

BH1	BH2
200 parts by weight of fresh cow manure	The same proportions and ingredients plus 0.2 g each of 502–507
23 parts by weight of straw	
33 parts by weight of earth	

This material was filled in sterile Blake bottles, rather loosely, with the broader side upward to give maximum aeration. Only sterile air was admitted. The bottles were stored in the dark. After ten days we already observed a complete transformation into a brown humus in BH2 (treated) case. The odor of manure had disappeared. Although the fibrous structure of the straw was still visible it too was transformed. The great surprise of this arrangement was the speed of rotting. The figures of the available minerals, especially of the nitrates, were also interesting. All the figures are in lb/ton; analysis was made after two months of fermentation. The figure in brackets indicates the number of analyses made, the table contains mean values. There is a certain fluctuation up to 2%, in the testing, due to variations in sampling. In case the first two values were not the same, the test was repeated in order to obtain more reliable averages.

Compound lb/ton*	BH1 without preparations	BH2 with 502–507
pH	8.0	8.0
Potassium	11.44 (2)	11.28 (3)
Calcium	7.48	7.48
Magnesium	0.2	0.2
Nitrate nitrogen	18.48 (4)	61.6 (6)
Ammonia nitrogen	trace	trace
Phosphates	14.1 (4)	14.8 (3)
Manganese	trace	low

* 1 lb per US (short) ton = 0.500 kg per (metric) tonne

The total bacteria count of BH1 after two months was 800 million bacteria per gram material, as determined by the plate counting method on beef-agar peptone at 29°C after 48 hours. The count of BH2 was 3 billion. Since the only difference between BH1 and BH2 was the addition of the preparations 502 to 507, this increase must be attributed to the effects of the preparations. All other conditions were exactly the same. The bottles were kept under sterile conditions in order to eliminate everything but the original materials used. These BH series represent ideal conditions until the organic matter present is digested. Then the bacteria count declines (beginning after three months) and in ten months has reached a trifle, 120 million (BH1) and 150 million. (BH2).

No top or bottom layer exists if the conditions are maintained so that aeration is possible throughout the manure. Fairly loose manure, not packed tightly, reacts best. However, too loose and too dry a manure heats up and no nitrogen is fixed. The proper limits, therefore, for a favorable fermentation are given. Again we repeat: a thin cover with good earth is essential to obtain the optimum and lasting result. Then we observe that the high count remains, even after ten months. This effect is particularly lasting if the earth has been treated with No 500, while the manure was treated with No 502 to No 507. Should this earth derive from good soil with earthworm castings in it, so much the better. Subsoil will not produce these results.

The anaerobic bottom and green layer of manure maintains a high level of anaerobic bacteria as the count of bottom layers in our bottle series shows, but at the expense of nitrogen fixation. Untreated manure increases, 1 month after being set up, to a count of 1 billion in the top layer but declines to 340 million after ten months. The bottom layer of this bottled manure started at 310 million, plus one species too numerous to count, but declined to 310 million after 10 months. Cow manure alone, plus No 502 to No 507, reached the peak of 1.4 billion/gram after three to four months but declined to 400 million after ten months. Earthworm castings in sterile bottles maintained the count of 1.15 to 1.8 billion up to ten months.

Recommendations

Based on these studies, our recommendations for practice are, to the best of our present knowledge:

1. Use manure which is spongy, moist and fairly well mixed with litter.
2. Set the heap up on bare topsoil.
3. Cover with good humus-topsoil. If the material is wet or packs tight sprinkle topsoil between the layers and build air channels with straw or brush and/or a bottom air drain.
4. Use preparations No 502 to No 507 inside the heap. Spray No 500 on the bottom earth and all earth to be used for interlayering and covering.
5. After six weeks the fermentation will be near its peak and will remain there for several months. The manure should be used after 3 months up to about 8 months if one wants to get maximum results.
6. Do not let the heap dry out or overheat.

We realize that we are only at the beginning of a rather new line of research. But from the little we have seen and the figures obtained we conclude that proper manure and compost making can be made a science like any other fermentation process which is harnessed for the benefit of humankind. It can even be considered to be an art. It is not a fad, there is no secret, no hocus-pocus, but it is foolishness not to make the best possible use of the hints nature herself has given to broaden our views and knowledge in regard to proper humus and compost treatment. So much is said nowadays about deficiency symptoms in soils and foods. Why not start right there where we can avoid wasteful practices and really make our fertilizer fertile?

To soil-conservation, manure, humus, and organic matter conservation should be added. We do not deny that deficiencies have to be remedied by adding what is deficient. But have our modern practices not increased deficiencies rather than taken care of them? The improper and unskillful handling of manure and organic wastes has contributed to create additional losses. Why not insert the lever at the source and reverse the wasteful practices? Here is capital which the farmer has on hand and does not need to buy, were he only to accept and learn how to use it instead of wasting it.

Our practical experience shows that with the organic method, supported by treatment of manure and compost with the biodynamic preparations and proper crop rotations, we can improve the soil at least in a few respects. My farm at Chester, N.Y., where I practice the biodynamic method, offers a few highlights in soil improvement. The soil there is a medium heavy clay with some lighter sandy clay interspersed, and lots of stones. Some fields are still in a poor state of drainage. It is a diversified dairy farm which provides from 200 to 250 tons of manure yearly, for an area of 135 to 150 acres (55–60 ha). Not all the fields have been completely converted as yet, only about 40%. About 30% are not converted and the balance is in process of conversion. Remarkable differences can be seen, first of all an improvement in pH. Four years ago we started out with an average pH below 6.0, mostly 5.5 to 6.0. Last year six fields were better than 6.0. This year 18 fields out of a total of 24 tested are better than 6.0, ten of them better than 6.5.

The nitrate content, without using any nitrogen fertilizer, just by preserving and stimulating the nitrogen fixation in the manure and on the fields, has increased in 11 fields, declined only in 2 fields (which were not treated). Increases have been observed as follows, in lb/acre:

No 1 from 16 to 20
No 3 from 16 to 24

No 5 from 12 to 16

No 6 80

No 7a from 12 to 80 (this field has had tile drainage
   installed and was fallow, volunteer clover has been the
   main "weed")

Field No 7b from 12 to 80

No 8 to 32

No 10 and No 11 were not converted as yet and declined
   from 12 to 4 and from 10 to 5

No 12 increased from 3 to 20

No 23 from 16 to 28.

We now have clover, which according to the previous owner
would not last more than 2 years, on the converted fields in the
fourth year and the stand is better than in the previous years.
Clover comes back, sown or volunteer, whenever we touch the soil
with the biodynamic treatment. Every visitor could easily spot the
converted fields as against those which have not been converted.
With all conservatism, we can say that no acidity or nitrogen
problems exist in the converted fields.

The organic matter content of our fields ranks as follows:

7 fields are better than 4%

10 fields 3 to 4%

3 fields 2 to 3%

only one is below 2% (namely 1.9%) and this one has not
   been touched with the biodynamic treatment yet.

Of course, a soil conserving crop rotation and proper tillage
are essential. These data are presented only as illustrations and
are not meant as complete proof. Conditions on a farm are ever
changing, many outside factors are involved too and it requires
years of observation before one can speak of "proof." The practic-
ing farmer, however, appreciates a turn for the better whenever it
occurs, inasmuch as he sees the jump in milk production every
time the cows go on an improved and converted pasture.

A little incident might illustrate the problem. We bought the

La Motte equipment for organic matter testing. At first the test as prescribed did not work. We asked the La Motte chemists for advice. It appeared that they had never had any complaints before. The amounts of testing materials, as indicated in the directions, were geared to organic matter below 3%. Evidently other users did not encounter difficulties because they never had soils with more than a 3% content. A simple adjustment of technique made it possible for us to use the equipment for our soils.

Questions and Answers on Biodynamics

What are the biodynamic principles?
To restore to the soil the organic matter which it needs so badly in order to *hold* its fertility in the form of the very best humus.

To restore to the soil a balanced system of functions. This requires our looking at the soil not only as a mixture or aggregation of chemicals, mineral or organic, but as a living system. We therefore speak of a *living soil*, including here both its microlife and the conditions under which this microlife can be fully established, maintained and increased.

While the biodynamic method does not deny the role and importance of the mineral constituents of the soil, especially the so-called fertilizer elements and compounds that include nitrogen, phosphate, potash, lime, magnesium and the trace minerals, it sponsors the most skillful use of organic matter as the basic factor for soil life.

However, the biodynamic method is more than just another organic method. It stands for a truly scientific way of producing humus. Not merely the application of "nothing but" organic matter in a more or less decomposed form is intended, but the use of the completely digested form of crude organic matter known as *stabilized, stable or lasting humus*. In this aim the method differs from what is commonly called "organic" farming. In the latter, any collection of any organic matter is apt to be called compost. In the biodynamic method the organic material to be used as a basis for compost is transformed either the biodynamic preparations, or by biodynamic compost starter.

It should not be forgotten that at the time of the creation of the method, in 1922–24, and afterwards during the years of research and experiments (from 1924 to about 1930), agriculture was dominated by the agricultural-chemical concept based on Justus von Liebig's research with regard to the major mineral fertilizer elements. A one-sided situation had developed. Nitrogen, phosphate, potash, lime were considered the only important fertilizers and trace minerals were ignored. Barnyard manure was looked down upon as an unimportant factor, frequently as a nuisance which had to be disposed of one way or another.

A fundamental change in the estimation of the value of manure and compost took place since 1930, increasingly since 1940, and of the trace elements since 1950. This went so far that manure and compost have now been restored to their proper, all-important position in modern agriculture, even in the orthodox school.

In biodynamics we are not only concerned with fertilizer but with the skillful application of all the factors contributing to soil life and health. So it is necessary to understand that life is more than just chemicals (inorganic and organic). Life and health depend on the interaction of matter and energies. A plant grows under the influence of light and warmth, that is, of *energies*, and it transforms these energies into chemically active energies by way of photosynthesis. A plant consists not only of mineral elements, or inorganic matter — these elements make up only 2–5% (in a few wild plants and weeds up to 10%) of its substance — but also of organic matter such as protein, carbohydrates, cellulose, starch, all of which derive from the air (carbon dioxide, nitrogen, oxygen) and make up the major part of the plant mass aside from water, namely 15–20%. The greater part of the plant mass, some 70% or more, consists of water.

The interaction of the substantial components and energy factors forms a balanced system. Only when a soil is balanced can a healthy, well balanced plant grow and transmit both substance and energy as food. We live not only from substance (matter), we also need energies (life-giving and life-maintaining). It is the aim of biodynamics to establish a system that brings into balance *all* factors which maintain life.

Were we to concentrate only on nitrogen, phosphate and potash, we would neglect the important role of biocatalysts (like the trace minerals), of enzymes, growth hormones and other transmitters of energy reactions. Already in 1924 Rudolf Steiner had called our attention to the important role of the finer elements (now called trace elements) in connection with health and proper physiological functioning. Today this is common knowledge. Enzymes and growth substances are likewise important. In the biodynamic way of treating manure and composts the knowledge of enzymatic, hormone and other factors is included.

In order to restore and maintain the balance in a soil a proper crop rotation is necessary. Soil-exhausting crops with heavy demands on fertilizing elements should alternate with neutral or even fertility-restoring crops — on the farm as well as in the garden, and even in the forest.

A soil which has been put to maximum effort, producing corn, potatoes, tomatoes, peppers and cabbage, for instance (all of them greedy crops), should have a rest period with restoring crops such as all the legumes. Temporary cover with grass and clover pastures helps to improve the humus and nitrogen situation. Crops that exhaust and arable cultivation consume humus. The soil must be given time to build it up again.

The biodynamic method therefore has emphasized the importance of crop rotation from its very beginnings. Cover crops and green manuring also play an important role in it.

The entire environment of a farm or garden is of importance too. It is obvious that polluted air loaded with the breakdown products of industrial and city combustion, gasoline and oil fumes, sulfuric acid, can be detrimental to plant growth. It is less obvious that many other environmental factors also affect the functioning of a biological system. Deforested hillsides are exposed to erosion. The water balance may be destroyed in such cases, the groundwater level dropping. The results of man-made deserts are only too well-known. To restore the most beneficial environmental conditions (forest, wind protection, water regulation), has been an important aim of biodynamics from its earliest years. Had the method been accepted before 1930, no soil conservation agencies would have been needed later in 1935 and the following years.

The soil is not only a chemical, mineral-organic system, but it also has a physical structure. The maintenance of a crumbly, friable, deep, well-aerated structure is absolutely essential if one wants to have fertile soil. All factors which lead to structural disintegration of the soil (like plowing of a too wet soil, and especially the deep plowing of wet clay soils) and what causes the formation of separating layers (hardpan), are things that have to be known. The biodynamic method is very specific about the proper cultivation of the soil in order to avoid structural damage. Many a farmer, even among the organic farmers, has defeated his aim by ruining the soil structure through unskilled cultivation.

Is biodymanics something for only a privileged few, or can it be used by everyone?
It has been said by persons outside biodynamic circles that this method represents the cream of organic farming principles. This does not, however, mean that it is restricted to a small group. It can be applied easily by anyone who cares to improve their handling of manure, composts, soil cultivation and crop rotation. Some of the steps to be taken are described below.

Build, and properly treat manure and compost piles. Do not waste any organic waste. Do not burn leaves and trash, but compost them. Collect everything. Do not apply crude, undecomposed organic matter to the fields or garden, but make use of the beneficial effects of microlife by first composting manure and all other organic material. Apply — immediately prior to planting or seeding — only predigested material which will not tie down nitrogen, phosphates and other fertilizer elements but will increase their availability. The use of the biodymanic compost preparations or starter will greatly help in reaching the goal: good humus.

Introduce soil-protecting crop rotations and cover crops.

Introduce green manuring, but take care that the green manure crop is properly plowed or disked under without tying down the soil life and nitrogen. In a garden, or wherever feasible, introduce mulching.

Improve your soil cultivation practices.

Establish proper environmental control, wind protection, good drainage, control of the water.

Is it a costly procedure to introduce the biodymanic method?
It is true that the building of compost piles requires some extra labor. However, this can be timed so that it does not interfere with the rush work on the farm. If one considers that by proper handling of manure and compost there will be no losses of soluble nutrients, the labor is well spent. If one considers further that the application rate of treated, well-rotted humus compost is less than that of fresh manure or crude compost, it is obvious that time otherwise spent in spreading and travelling over the fields is saved at the moment when time counts most.

In the long run the extra labour and expense for composting is well spent and will be returned in savings of nutrients and of time at other phases of the farm and garden work, including even the need of less cultivation since one gets a more friable humus soil.

The fertilizer value of manure and compost can be considerably increased by the biodynamic method. Humus-building techniques also will help the fertilizer effects to last longer.

Is it possible to make a farm entirely self-suffient with regard to fertilizer elements, or does one still need to buy supplements from the outside?
These questions can be answered only in each specific instance. If

there are deficiencies, they must be taken care of. However, applying biodynamic methods enables the farmer to reduce deficiencies to the minimum. Humus deficiency is the most important one, because without humus one is not able to hold and build up a soil. This needs to be taken care of first of all. A soil below 1.5% organic matter is only on a maintenance level. A soil above 2% organic matter begins to build up reserves. Only when the level of continuing existence is reached can one tell how much else is needed.

Many times we find hidden reserves, which need only to be made available. In mineralized soils there are no more reserves.

The answer to the above question depends also on the crop rotation. Corn-wheat, corn-wheat, corn-wheat, over a long period of years can only exhaust a soil, no matter how much organic matter is applied. Soil conserving or protecting years, between exhausting years, are an absolute necessity.

On many biodynamic farms the problem of self-sufficiency has been solved in practice.

Does the application of the biodynamic method require special studies or efforts?

If somebody is a good practical farmer or gardener he might just as well become a biodynamic farmer or gardener, with the slight added effort indicated above. If he is not up to standard he will have to improve his farming methods. This he must do anyhow if he expects to continue with any degree of success, and he should not think that by using the biodynamic method, or for that matter any organic method, he can escape the necessity of improving his general practices.

Expert advice is available through Biodynamic Associations (see Useful Contacts at end of book), to show any farmer or gardener how to start improving his practices from the biodynamic point of view. The first step is always to make a survey and to take inventory of your particular situation in order to plan intelligently.

Why, then, is biodynamic farming still so little known and practiced?
Biodynamic methods were well known to those who opposed
them — in the Liebig fertilizer camp. To them they must have
appeared a real danger. This was in the 1920s and 1930s. Now
the understanding for biological balance and organic principles is
common knowledge and generally accepted.

There is, however, a reason why biodynamics did not spread.
It is based on human nature and is not an agricultural problem.
It is about the most difficult thing imaginable for human beings
to change old habits, old customs, and (in this context) to begin
to think in terms of biological balance, soil life and health, rather
than of NPK only.

Biodynamics has no single recipe to offer, but requires some
coordination of farm planning on a long-range program. Then
there is the fact that many farmers think only in terms of quantity
yield, not quality. Only such methods are put to use which prom-
ise bigger yields.

The yield depends on many factors beyond the control of biodynamics: water supply, too much rain or drought, seed quality,
and above all the farmer himself. What we *can* do is show that
soil improvements *have* been obtained, that biodynamic farmers
have very little trouble with livestock and plant diseases (especially
having no breeding or sterility trouble), are not bothered with
lodging (plants falling over because of weak stems) in wet years,
produce crops with maximum protein and vitamin contents. This
we know: we get the top quality which can be produced. Also,
the quantity yield of a good biodynamic farmer has always stood
above the average level.

The introduction of the biodynamic method therefore goes
hand in hand with a striving for better quality. When there is
interest in a better quality of food and feedstuffs, biodynamics
is in its proper place. Health conscious people everywhere have
always been asking for, and been appreciative of, biodynamic
products.

Does biodynamic farming avoid or make unnecessary the use of poisonous sprays against insect pests?

Biodynamics does not completely counteract insect pests. To make such a claim would create a false impression. The important question is not, "Are insect pests present?" but rather, "Do they spread out, and do they produce measurable damage?"

A few insects may be present, this is always possible. They may be wind-borne, or move in from infected areas. That is bound to happen occasionally. But we have found, in our thirty years of biodynamic farming experience, that they definitely did not spread or do great economic damage.

The question of insect pests is one of biological balance and control. Poisoning sprays have not solved, nor can they solve, the problem. If the biological balance is restored the situation will be entirely different.

Ehrenfried Pfeiffer

by Herbert H. Koepf

Ehrenfried Pfeiffer holds a special position among the personalities who from 1924 onwards have developed the biodynamic method of agriculture. As a researcher and adviser he was in the forefront; yet he maintained a keen awareness of biodynamic farmers' practical needs. Out of spiritual insight he traveled new roads in natural science; the course of his life exemplifies human work that springs from a spiritual impulse.

The early years

Ehrenfried Pfeiffer was born in Munich, Germany on Sunday, February 19, 1899. At the age of five he lost his beloved father, and the Pfeiffer family moved to Nuremberg where young Pfeiffer's maternal grandmother provided the child the warmth and harmony which all children need. Pfeiffer's maternal grandfather, who was a renowned apothecary, introduced him to medicinal herbs, taking the boy to his laboratory to show him chemical reactions and thus awakening the love for natural phenomena which became so important for Pfeiffer throughout his life. As a schoolboy, Ehrenfried liked to be by himself in woods and fields, as both then and later, he was familiar with the elemental beings.

At about age eight, as the sensitive, sometimes effervescent nature of the boy stabilized, he began to distinguish himself among his fellow pupils and impress his teachers. A second stream in his life revealed Pfeiffer's intimate understanding of his human and natural environment. Pfeiffer demonstrated musical talent:

he had absolute pitch, and for a time he was sent to the musical conservatory. His sensitive hearing deepened his perception of voices, of sounds in nature and in the cosmic expanses. When he was fourteen, his mother entered a second marriage to Theodor Binder, an economist, who became a mentor for Pfeiffer. During the third seven-year period of his life, Pfeiffer decided on the direction of his life: he devoted himself to natural science and, initially, also to technology.

Before beginning his career, however, he joined a sappers unit during the last phase of the First World War. At one time, when in immediate danger, Pfeiffer received an intuition that he would be spared for later tasks. In a lecture which he gave in September 1961 — two months before he passed away — Pfeiffer described how during his youth he turned to nature:

> When I was fourteen or fifteen years of age, I had
> the idea, the ideal, to grasp the forces which are
> working behind the natural phenomena, which make
> plants grow, which give mobility to animals, produce
> lightning and thunder, and which are active in the
> functions of my body as well. I wanted to search for
> that realm of forces which gives agility to everything.
> Such wishful dreams induced me to study our
> environment, plants, animals, and nature's forces.
> These juvenile impulses remained imprinted in my
> whole rich life as a researcher.

While in Stuttgart shortly after the war, Pfeiffer met Rudolf Steiner, who was then lecturing on social renewal to a large assembly of workmen. This meeting determined the further course of Pfeiffer's life.

In 1919, Steiner entrusted Pfeiffer's stepfather, Theodor Binder, with the direction of the economic aspects of the continuing construction of the Goetheanum in Dornach, Switzerland. The family moved to Dornach while the building's further design and construction were under the artistic leadership of Rudolf Steiner. The big structure with its two cupolas, its huge columns, its architraves and carved stained glass windows had been erected during the First World War as citizens of various nations, even those at war with each other, joined in the work. From 1919 onwards the building, though not quite completed, was put to use. At the time, Pfeiffer, still reading physics and electric engineering in Stuttgart, was invited to visit Dornach, and before long was asked to join the work. Using his own novel design, Pfeiffer constructed the lighting for a stage in a carpenter's workshop adjacent to the Goetheanum. Later, for a stage in the Goetheanum itself, lighting of similar design was used to satisfy Steiner's indications requiring a modulation of colors which could be an integral part of what happens on the stage.

In 1920, as a result of his visit, the twenty-one-year-old Pfeiffer moved to Dornach, So between 1920 and 1925, Pfeiffer stayed close to Rudolf Steiner, studying under Steiner's continued guidance, and living near the Goetheanum, which became the center of Steiner's work for the further development of anthroposophy. During this period, Pfeiffer lived in the midst of tremendous developments in the anthroposophical movement.

Before continuing, let us briefly review the burgeoning anthroposophical movement at the beginning of the twentieth century, a milieu in which the young Ehrenfried Pfeiffer began to entertain ideas for his own research projects.

From the beginning of the 1880s, Rudolf Steiner became known for his philosophical and editorial work. Steiner had edited Goethe's scientific writings and published his own major books on philosophy, in particular his *Philosophy of Spiritual Activity* which appeared in 1894. From 1901 Steiner lectured publicly on his spiritual knowledge, initially within the Theosophical Society, and from 1912 onwards in the Anthroposophical Society that he then founded.

As a scientist of the spirit, Steiner spoke lucidly, allowing the human consciousness in this age of natural science to begin to comprehend spiritual matters. Steiner's lectures fostered under-standing of the tripartite, body/soul/spirit nature of the human being, revealed insights in the evolution of humankind and the kingdoms of nature, and suggested a method by which such spiritual-scientific knowledge can be attained. The Goetheanum itself marked a new departure in architecture and other visual and performing arts. A new art of movement, eurythmy, and new knowledge of speech formation proceeded from the Goetheanum.

With the opening of the first Waldorf School in 1919 in Stuttgart, a new pedagogy presented itself to the world. The work of Steiner and his close associates showed new approaches to medicine, to pharmacology, and to curative education. In 1917 presentations on a new social order had begun. Most significantly for Pfeiffer, 1924 saw the birth of a new agriculture. As a modern spir-itual impulse, anthroposophy pointed the way to the kind of cultural renewal which was needed at the time and is still needed now.

From the beginning of the modern age, from about from the fifteenth century onwards, the development of the individual per-sonality, its world view and drive for freedom, have been nurtured by a new mode of thinking, one developed by natural science and demanding that thought proceed from exact observation of nature. The situation of the present demonstrates convincingly the theoretical and practical limitations of materialistic thought. Nonetheless, despite its emphasis on observation of nature, this mode of natural-scientific thinking does not necessarily result in narrow materialism. Indeed, the faculty for clear thinking and sound judgement thus attained are prerequisites for understand-ing anthroposophy. And Rudolf Steiner recognized in the young Ehrenfried Pfeiffer an individual capable of such thinking, but one who might contribute to overcome materialism in science.

Following Steiner's advice, Pfeiffer chose to continue his stud-ies in Basle, electing to major in chemistry. Pfeiffer was advised above all to register for laboratory courses, but also to survey mineralogy, physical chemistry, botany, plant ecology and general ecology — even to look into economy, sociology and psychology. In 1955 Ehrenfried Pfeiffer reported a conversation with Rudolf Steiner which explains the logic of the advice that Pfeiffer study the natural sciences:

> If you want to overcome the materialistic world conception, then the first thing is to know this materialistic world view, understand it, and then you can overcome it. You must seize this bull by its horns and disprove materialistic science, using its own methods. This for me has become the salient point.

Steiner further advised Pfeiffer that his primary task was to collect observations of phenomena, that interpretations are of lesser importance. Pfeiffer's notebooks provide evidence that he took Steiner's advice in earnest. On one page of his lecture book Pfeiffer noted phenomena and how they are explained; on the opposite page he noted how to understand these phenomena from the perspective of spiritual science. Pfeiffer also compiled material on the many objects and personalities which Steiner mentioned in his lectures. Clearly both Pfeiffer and Steiner held one another in high regard: Steiner called Pfeiffer by his Christian name, and Pfeiffer recognized Steiner's greatness and considered Steiner to be his spiritual teacher.

Pfeiffer asked questions. He was interested in the realm of forces which are active in the life of the earth and in human beings, asking how the etheric formative forces can ostensibly be demonstrated and on the whole be directed into a manner that one can experiment with them? How would it be possible to make use in a technology of the etheric as a natural force, for it is in the nature of the etheric not to destroy but rather to build, and thus might a technology which builds up evolve? (Pfeiffer on February 27, 1955).

Steiner suggested organisms (such as wheat seedlings) which when used as reagents can make visible rhythms and formative forces in the processes of life. Not until years later would research in rhythms be taken up, and today chronobiology is a major field in biology.

These were the questions which were first posed in 1920–21. Under the most modest of circumstances experimental work was begun, one result of which was the method of sensitive crystallization. In 1921, a research laboratory was established at the Goetheanum, and for many years Pfeiffer and his coworkers were responsible for the work in biology.

Dr Guenther Wachsmuth, who was appointed leader of the science section, reports on his joint work with Pfeiffer, who sought means by which his experimental work could be made useful for agriculture. In 1922, Steiner for the first time indicated how to make those substances which later became known as the biodynamic preparations. These preparations consist of certain substances which are taken from the plant and animal kingdoms and exposed to seasonal, environmental influences. Minute quantities of these substances are then applied to soils, plants, and farm-produced organic fertilizers. Steiner personally supervised these experiments, and was present when the first preparation was taken out of the soil. Wachsmuth relates how Steiner stirred and applied it in the field. Finally in 1924, Rudolf Steiner gave his eight lectures for farmers in Koberwitz, Silesia, describing in detail the preparations which are an essential part of the biodynamic method of agriculture.

In addition to his research, Pfeiffer was much in demand for other tasks. He traveled with Steiner on lecture tours, and such times of travel provided Pfeiffer opportunity to ask Steiner questions. Pfeiffer gave lectures himself in Switzerland and was assigned to guide tours through the Goetheanum, helping visitors become aware of the language which one could discover in the building's architectural and sculptural forms. Steiner made certain that Pfeiffer was present when he gave courses to medical doctors; which otherwise were exclusively for medical specialists. Throughout his acquaintance with Steiner, Pfeiffer was a lecturer on agriculture, but missed the 1924 lecture course in Koberwitz, as he was asked to remain in Dornach to care for a workman who had suffered an accident during the construction work.

During the night of New Year's Eve 1922–23, the first Goetheanum was destroyed by fire. The building had been the fruit of ten years of work. Evidence from the fire suggested arson: the flames spread so quickly through the structure that the efforts of the fire brigades which rushed to the site were without success. Pfeiffer remained throughout the night at Rudolf Steiner's side, for after having done everything possible to providing evidence on the causes, Steiner remained with others for a long time gazing at the huge sea of flames. This event connected Pfeiffer yet more deeply with his fellow man and his teacher, Rudolf Steiner. We can see something of the power of the spiritual impulse and the nature of the people carrying them, when we realize that after this tragic night, the proceedings of the conference in progress at the time, continued the following day without interruption and on schedule in improvised spaces. The continuity of the spiritual task was affirmed, in spite of a disaster of such dimensions.

In spite of the fire, the work of Steiner and his researchers continued with redoubled vigor. However, on September 28, 1924, illness forced Steiner to stop lecturing. There followed six months during which Steiner lay on his sick bed, until March 30, 1925, when, despite much devoted medical care, Steiner cast off his earthly sheath. During that time, Pfeiffer remained in Dornach and with the others present shared the grief of their teacher's illness and death.

The period of Pfeiffer's learning from 1920 until 1925 had its inner aspect as well. About this Alla Selawry remarks that,

> these tasks need a sort of spiritual schooling in
> which two capacities must be exercised simultaneously:
> the first is a love of nature in all its details, precise
> observation, a careful study of phenomena; the second

is a practical work with the forces of nature, a genuine
knowledge of substances, an intuition in standing at the
altar of nature, that is the true Rosicrucian impulse for
the transmutation of substances.

A few paragraphs later she remarks:

In particular Ehrenfried Pfeiffer used all his gifts
to further the spiritual aims of anthroposophy.
As a pupil of the spirit he endeavored to develop
the organs of spiritual perception. As a developing
scientist and researcher he sought to translate
spiritual insight into experimental arrangements and
to fructify science with spiritual science.

Pfeiffer as scientist and practitioner

These then are the tasks to which Pfeiffer devoted all of his
strength. At the Goetheanum and in the research laboratory he
developed his activities as an experimenter and a teacher. Pfeiffer
concerned himself with biodynamic agriculture, nutritional quality, medical problems, and the method of sensitive crystallization.
Soon a number of young individuals joined him to work under
his guidance, often with a minimum of equipment. In the Youth
School which was in progress at the time, he taught botany, zool-
ogy and chemistry, always offering his students spiritual insight.
Then, as throughout his life, despite low funding, Pfeiffer was
able to manage a considerable number of projects because faithful
coworkers helped selflessly.

A pioneer, alert to current issues in both practical affairs and
scientific studies, he identified topical issues and immediately
took them on. Even now, decades later, one still encounters people
whose enthusiasm for the cause is as vital now as long ago, for
those who came into contact with Pfeiffer's enthusiasm retained it
throughout their lives. In the laboratory and in the greenhouse he

instilled in his charges utmost conscientiousness: even temporary helpers, frequently just instructed for some specific task at hand, learned from him to carry out their work reliably and responsibly.

Geographically the scope of his work expanded throughout Europe and soon to America. Lectures, advisory work, and visits to biodynamic farms took him from northern Europe to the Mediterranean, even to Sicily and Egypt. Often Pfeiffer was able to accomplish the task of his trip and still have time for a visit to important sites and mystery centers of ancient epochs, where he could trace the still lingering sense of elementary life.

The growing biodynamic agricultural movement is indebted to him for much more than his many lectures and advisory tours: Pfeiffer made many scientific contributions, broadening our knowledge of the nutritional quality of foodstuffs, the development, use, and effects of biodynamic preparations, companion planting, and other subtle effects on plant growth. His suggestions for agriculture fit the practical needs for sustainable land use. The first edition of his book, *Soil Fertility, Its Renewal And Preservation*, appeared in German in 1938. This work was to become the textbook on the biodynamic method in agriculture and has been translated in a number of languages. It summarizes research and techniques developed in Europe and America. The book discusses crop rotation, crop management, and treats composting and other means of fertility extensively. Promoting the idea of the farm organized and managed in the image of an organism, the book includes a chapter on the importance of the family farm for the health and preservation of the rural areas, a subject which is not to be underestimated.

A Dutch woman, Maria Tak van Poortvliet, had attended Pfeiffer's classes in the Youth School. Around 1927–28 she handed over to Pfeiffer all her land holdings in Holland, a total of 568 acres (230 ha) divided in several farms which were consolidated in the Loverendale Cultural Inc. The majority of the shares were given to Pfeiffer. Of his own initiative Pfeiffer introduced the biodynamic method and, linking up with the knowledge already there in the country, he further developed and convincingly demonstrated the biodynamic method. In 1930 vegetable growing began on 17.5 acres (7 ha). A slowly rotating mill for careful processing of grain and a bakery was established and a distribution service delivered produce throughout the country. The quality produce and the courses on nutrition which Pfeiffer offered stimulated the public's interest in quality foodstuffs based on a holistic method of production. In spite of difficulties which had arisen during the world economic crisis of 1929, the Dutch biodynamic farms prospered. Later, when Pfeiffer was in need of money for his work in America, he did not need to borrow funds. The Loverendale farms provided much impetus for the growth of the biodynamic association in Holland and later on for founding the Warmonder Hof Agricultural and Horticultural School. The produce from the Loverendale enterprises are cherished to this day.

The sensitive crystallization method

Earlier, mention was made of Pfeiffer's search for experiments to show the effects of etheric formative forces. In 1930 Pfeiffer wrote about this topic in the yearbook *Gaia Sophia:*

> Rudolf Steiner's suggestion led to a search for a
> reagent on etheric effects: that is to discover a natural
> process which reacts to etheric formative forces so
> subtly that in every case after the reaction different
> forms result, depending on specific influence. In order
> to find this reagent one should observe the processes
> of crystallization influenced by plant substances or
> blood, and study what changes are brought about. "I
> can't tell you yet what you will discover; in any case
> you will be surprised how much you will find." This
> is all that Rudolf Steiner said about the subject, and

what one had to start with. When I tried to inquire
about the experimental arrangement, he repeatedly
said: "This is what you have to find yourself."

Pfeiffer then chose a horizontal planar plate, which allows the
crystals to spread. It was Erica Sabarth who from a number of
salts took copper chloride dihydrate ($CuCl_2.H_2O$), which proved
to be suitable. Together with her, Pfeiffer then worked out the
method. Trained as a chemist, Sabarth became Pfeiffer's lifelong
faithful co-worker in Europe, North America and for a time in
Central America.

The method turned out to have potential. If copper chloride
is allowed to crystallize under controlled conditions, one will
find on the plate an unorganized spreading of the fine crystalline
needles typical for the salt. If, however, one adds to the process a
tiny quantity of blood, some other body fluid, or the water extract
of some vegetable substance, then, surprisingly, the crystals form
into a coordinated pattern. Proceeding from one or several centers a coordinated pattern of branched bundles of needles radiates
towards the periphery of the glass disc; sometimes the pattern
includes hollow, rounded fields. The pattern is specific for the
substance tested and can be reproduced. On the whole, the pat-
terns which are brought about by different substances show a
relatively broad spectrum of forms. They are evaluated by describing the texture and discerning between specific textural types. A.
Neuhaus remarks that when applied, the organic substance itself
does not enter in the crystals. Apart from temperature and other
external parameters, the substance influences the pattern from without. One evaluates the total picture and the detailed characteristics
of the crystallization. After the ground-breaking work of Pfeiffer
and Sabarth, others were able to take up the method, including
from among the many whose names could be given here, Frieda
Bessenich, Magda Engqvist, Hans Krüger and Alla Selawry.

Sensitive crystallization shows one way to an understanding

of the formative forces in organic life. It is only a salt and some organic substance which are present on the plate. As organic substance is fashioned by life, it, too, might be analyzed. By such analysis one may find data which can be interpreted according to existing theoretical knowledge, or if the theory does not fit, it might be adjusted. Yet what is revealed by the crystallization is not an inference per se; rather, the statement evolves from observation and comparison with an established, ordered body of observations and experience. One must learn to read the language of the "picture" by relating both its overall appearance and individual characteristics to the kind and origin of the sample. During 1938–39 Pfeiffer and Sabarth accepted an invitation to the Hahnemann Medical College in Philadelphia, Pennsylvania, to install a laboratory for crystallization and to evaluate tests on blood samples taken from patients suffering from cancer. In 1939, Pfeiffer wrote:

> Our work here went quite well. We limited ourselves
> to a statistical evaluation of clinically well-established
> pathological symptoms. Over one thousand such
> cases were studied in a little over a year, all verified
> histologically-pathologically. Definite indication
> for cancer positive 83%, for cancer negative 93%.
> Now scientific objections are no longer possible. In
> the future all will depend on how intelligently and
> faultlessly the work is performed.

This work earned Pfeiffer an honorary doctorate from Hahnemann Medical College. Today, laboratories in a number of countries make the crystallization method available for medical diagnosis.

The original question regarding formative forces suggests that vegetable substances may be regarded as suitable samples for the crystallization method. To the independent view the plant and its ontogenesis make visible two levels of existence: matter and life. The observer perceives the form of the plant, its organs, and their material configuration as they undergo transformation in time. The genetic makeup, general, and chemical conditions of growth of the plant are only the physical aspects of an organism. This specific appearance becomes manifest under the influence of the organism's earthly and cosmic environment, which we recognize in the typical characteristics of a landscape and the cycle of the year. The crystallization picture made of water extracts of vegetable matter adds additional features to the essential image of the specific plant, or a specific species, as it becomes manifest in form and growth.

The method of sensitive crystallization is frequently used as a qualitative test for the potential shelf life of foodstuffs. Depending on the specific questions, the method may be employed in conjunction with data from chemical analysis, or without such analysis as well. To just mention a few examples: samples of flour of different origin and degree of milling (fineness) have been tested, as has milk processed in different ways. Definite variations of the pattern of crystals are achieved by plant extracts derived from samples which vary according to stage of growth, ripeness, and, above all, shelf life of perishables. The process can demonstrate the effects on plant growth from such factors as soil type, fertilization, and biodynamic treatment.

When used to test progress of plant extract decomposition over time, sensitive crystallization can indicate whether or not ripening and balanced environmental conditions prevailed during plant growth. Biologists investigate the formative potential in various plant organs or tissue that is alive, stable, or beginning to die. Such investigations may be made for two ends: as a quick qualitative test for an initial orientation, or for analysis of the morphology of finer characteristics which are not revealed by a number of analytical data. In the course of time, a broad field of uses for the method has evolved.

At a later point the designation "picture creating methods" was used for sensitive crystallization and a few other methods. These are the chromatogram as devised by Pfeiffer and also the capillary method of Lily Kolisko (also called capillary dynamolysis) and the "drop method" which has been developed by Theodor Schwenk. Pfeiffer's chromatography method, which is not difficult to carry out, is well-suited for investigating soils, compost, and other farm-produced organic fertilizers — in short for vegetable and animal wastes in the process of decomposition or humification. Testing laboratories which are mainly working for organic growers routinely use these chromatograms in their programs. However, both chromatograms and capillary dynamolysis are also used for testing food and medical plants. All these methods are at the root of the holistic search to make perceptible the formative and transformative processes in living entities.

Pfeiffer in the United States

In 1926 Charlotte Parker and her friends acquired the land of the Threefold Farm in Spring Valley, New York, about 35 miles northwest of New York City. In the course of time an anthroposophical residential and conference center emerged, with a guest house, school, and a number of cultural pursuits. This is the place where from the mid-1940s onwards the work of Pfeiffer and his coworkers found a home.

Biodynamic work in the States commenced soon after Steiner's lecture course on agriculture. During 1925–26 the first biodynamic preparations were made in a garden in Princeton, New Jersey. From the end of the 1920s until the mid-1930s, quite a number of people who belonged to the circle of Threefold Farm traveled to the Goetheanum, some also to Holland, to learn about the new method of agriculture. From 1932 to 1939 Pfeiffer made four or five visits to the United States, primarily attending the summer conferences at Threefold Farm and also Sunrise Farm in Maine, where for some time there was a laboratory staffed by one of Pfeiffer's coworkers. From 1948 until 1980 the agricultural summer conference of Threefold Farm was a regular annual event.

In 1930 biodynamic activities in the United States were in progress in a dozen states: the Bio-Dynamic Farming and Gardening Association was founded and incorporated in New York in 1938, and by 1939 between thirty and forty farms and a like number of gardens followed the biodynamic method. *Short Practical Instructions in the Use of the Biodynamic Method of Agriculture** (1935) by Pfeiffer and the translation of his textbook *Soil Fertility, Renewal and Preservation* (1938) provided the urgently needed literature. During the fourth biodynamic conference in 1939 Pfeiffer spoke about the need for a school for biodynamic agriculture.

H.A.W. Myrin and his wife Mabel Pew Myrin of the Sun Oil Company, owned 838 acres (339 ha) of agricultural land in Kimberton, Pennsylvania, about 40 miles west of Philadelphia. He met Pfeiffer, and because both shared an interest in biodynamic farming, in 1940 he invited Pfeiffer and his family to move to Kimberton. On the two dairy farms, which also served for development and demonstration, a farm school was established for several years. Thirty to forty students assembled for a theoretical and practical one-year course. In addition to this, winter courses of two to six weeks' duration took place from 1941–43, with lectures, laboratory practice, and field instructions. About this Pfeiffer said:

> The intention is to educate the practical farm
> manager rather than the textbook scientist. However,
> basic sciences are necessary to enable the farmer
> to work with full consciousness. But all fields of
> knowledge will be treated with a view to the daily
> practical problems. The goal is not to educate a
> scientist, but to enable a young farmer to become
> an independent and skilled agriculturalist and to
> learn how to combine soil conserving methods with
> extensive economic farming.

* Reprinted later under the title *Using the Biodynamic Compost Preparations and Sprays.*

Cropping, dairy farming, soil conservation, composting, crop rotations, greenhouse cultivation, beekeeping were all important components of the program. Hundreds of students, teachers, lecturers, local farmers and other visitors — among them J.I. Rodale and Paul Keene — frequently came to Kimberton.

The legacy of Kimberton Farms has endured: during the 1960s and 70s, one could frequently meet farmers in the States who had studied at the school in Kimberton. Still a long time after their study at Kimberton, these farmers spoke enthusiastically of the formative influence of Pfeiffer, the same tribute one hears as well from others who have worked with him. What they had learned from him was useful for a lifetime. However, as often happens between strong personalities, beginning in 1944 a small matter made further cooperation between Myrin and Pfeiffer impossible: Pfeiffer had to look for another place for his work. On an earlier occasion he had remarked: "If I should ever have to leave here, I shall buy a farm of my own and demonstrate that biodynamics is not a rich man's hobby." This is the step he now took.

At the beginning of March 1944 after two months of exploring, actively supported by Peter Escher, Pfeiffer found in Chester, New York a 285 acre (115 ha) dairy farm with 100 acres (40 ha) of tillable land. Students could not be accommodated there, but the farm was suited for development and research. Pfeiffer and his associates found the first summer there difficult, with a severe drought. Pfeiffer worked hard, for he had endured much during the past years. Indeed, he overworked himself, and as a result fell seriously ill: Pfeiffer had to spend two years in the hospital in Pomona, followed by an extended time of convalescence. His wife had to shoulder the burden of carrying on with the farm and bringing up the children.

After being released from the hospital Pfeiffer found an abode at Threefold Farm, at first in a temporary place provided for him to carry on with his work and later on the ground floor of the auditorium, which had been built for the conferences. There he occupied several rooms, establishing a laboratory for himself and his coworkers who came over from Europe. In addition to serving for lectures and education, the laboratory was used for investigating soils, composts, food stuffs and the like. Sensitive crystallization was made available to doctors. The finances remained small indeed; income had to be acquired through work and new initiatives. Throughout the country Pfeiffer was a well-known and a much sought after speaker on agriculture and nutrition. He was a co-founder and speaker for the Natural Food Associates (NFA), which promotes natural methods of food production without chemicals for soils, plants, or food processing. Finally, Fairleigh Dickinson University in Rutherford, New Jersey offered Pfeiffer a Professorship in Nutrition, which he attended to during the last decade of his life.

From the outset of his career, the quality of food, and the influence of soils and other environmental forces on produce quality were important themes in Pfeiffer's work. Teaching obligations offered Pfeiffer opportunities to further his field by study and experiments. Pfeiffer and his faithful staff accomplished not just routine laboratory work, but they produced much research on substances and their forces and virtues. Processes in and transformation of substances were pivotal themes in Pfeiffer's work. He studied the way vapors which contain biodynamic preparations and other substances influence root morphology. One of the last projects which he took up was the influence of heat treatment on particular changes in the protein of wheat seeds. Pfeiffer passed away at Threefold Farm on November 30, 1961.

Pfeiffer, researcher and teacher

Pfeiffer's mode of teaching shows characteristic features. Recognizing that spiritual insights opened new and independent roads for research and practice in agriculture, nutrition and medicine, Pfeiffer carefully adhered to true, that is not inferred, facts and to proper contemporary research methods. He allowed the attitude of the conscious soul to inform his advisory and public activities; he described factually what the immediate situation called for, leaving to the free will of others whether or not they chose to take an interest in the spiritual aspects of the issue. Apart from the biodynamic textbook, *Soil Fertility, Its Renewal and Preservation*, during the difficult times of the Second World War in Europe, Pfeiffer collaborated with E. Riese to publish a small guide for gardeners, *Grow a Garden and be Self Sufficient*.

From the mid-1930s on, concerns about soil erosion and soil conservation took on worldwide significance, especially given the example of the dust bowl in the southwestern States. Pfeiffer's 1947 book *The Earth's Face and Human Destiny*, which addressed the erosion problem and care for the landscape, is read to this day in many countries. For the biodynamic movement, he wrote a number of shorter pamphlets. From 1941 onwards, the quarterly *Biodynamics* appeared regularly, always including an article by Pfeiffer. Meeting widespread approval, these articles show the peculiarity that they don't really grow outdated even as methods of cultivating the land change and new knowledge is added. Pfeiffer's presentations and diction gave them a liveliness which keeps the reader alert and interested. His books are conceived with a profound understanding of the lasting principles which make for a thriving, sustainable agriculture. Indeed, what Pfeiffer had to say about the family farm is still of topical interest. Hence to this day a number of the Pfeiffer's titles are still in print.

However, being at once both practical and at the forefront of knowledge is not just a theme in Pfeiffer's writings. After having taken over his own farm in Chester, he did not miss the opportunity to monitor the progress of the farm as it was changed over to the new kind of management. In 1954 he published a detailed study on the seven years of the nutrient regime in the fields of the Chester farm. As he obtained the flame photometric apparatus for quickly detecting the presence of trace elements, he produced a long and detailed study of the biodynamic preparations. In his laboratory, Pfeiffer introduced paper chromatography for determining amino acids, and he developed a method for determining amino acids in urine, which process, together with sensitive crystallization, is most useful for physicians.

Even during his stay in the hospital, Pfeiffer found things to do: he worked in the laboratory, becoming familiar with microbiological methods. In the 1949 winter issue of *Bio-Dynamics* he announced the Bio-Dynamic Compost Starter which, "has been developed for use with grinding machines for rapid decomposition of garbage, weeds, leaves, municipal and industrial wastes." B-D Compost Starter is a carefully made ground compost using bacteria, biodynamic preparations and other additives.

Pfeiffer also turned his attention to refuse composting. Composting and disposal of garbage were — and still are — matters of great concern. During the 1950s larger municipalities lacked possibilities for getting rid of the rapidly increasing quantities of municipal waste. At that time the percentage of compostables in municipal wastes was high; only in the following years did the share of cardboard, synthetic materials, and toxic substances increase. Yet long before the necessity was commonly acknowledged, recycling was an important principle in the biodynamic method of farming. For a short while a composting plant operated in California. In the mid-1950s a fully equipped large-scale pilot project was in progress in Erlangen, Germany.

Because of his achievements in refuse composting, Pfeiffer received many invitations to lecture and consult on the subject, both in the United States and overseas. Pfeiffer traveled to the Azores, to Cuba and Taiwan. However, further spreading of the composting method of municipal waste disposal was limited by the fast growing quantities of raw material that was hardly suitable for composting and administrators' interest in instant and cheap solutions. Now, of course, there is renewed interest in composting of garbage.

Pfeiffer's spiritual quality shines through in the objectivity of his lectures and advisory work. When in 1957 the German edition of his compost manual was published, a friend remarked that the book did not quite do justice to the "organic" point of view, that composting NPK (nitrogen, phosphorus, and potassium) received too much attention. Pfeiffer replied in a letter, "Of course I must justify what I do with my own conscience." He emphasized that in all that he did, he was entirely consistent with the basic biodynamic concepts. (With this he pointed to the farm, which should as much as possible be conceived as a closed system, an organism, in which organic farm-produced manures are of foremost importance.) He added, however, that "whenever facts are established by experience and experiments, one cannot reject them. There is less need to concern oneself with theory."

He emphasized that shortcomings or weaknesses of the results of materialistic research should be exposed using scientific methods. This approach was rather more easily accepted in the United States than in Germany, where on earlier occasion Pfeiffer was faced with an over-emphasis on doctrine. But then he mentioned the "main point" which he explained by referring to the "law of the minimum," which law not only holds for the narrow NPK aspect, but, rather, in the widest sense, for every factor which, when present at a minimum rate, determines the functioning of the whole organism. The unit under consideration might be the soil, a plant, or a human being, and their environment, embracing the physical-material and the cosmic environment.

"All the nutrients must be included, even the trace elements, water, air, carbon dioxide, biologically important substances. One should also include energies — not just physical ones, but those in light, in the etheric, and in the astral realms, and also heredity which is but a modified image of the primeval cosmic order, etc." He also remarked on the state of balance, which nature always strives for, the antithesis between polar opposites out of which life is perpetually renewed. To act in accordance with this principle of continued regeneration is a guiding light for the active human being in his relation to nature.

Now and then one hears the opinion that in his agricultural and nutritional work, Pfeiffer was reserved regarding biodynamic concepts and anthroposophical ideas. Sometimes he is considered to actually have been a spokesman and researcher for "organic" agriculture, one who worked in the ways established by traditional scientific methods. Considered superficially, his writings and talks might occasionally have encouraged this impression. But he who stops at a superficial understanding knows little of Pfeiffer's work. Pfeiffer's reply to a factual question was always factual. He might have presented results of his own work and valid results of general contemporary research as well. He offers what one can expect from a good advisor: a practical suggestion which can be applied as such. He did not offer theoretical deliberations. This manner of Pfeiffer's characterizes the attitude of the consciousness soul. It conforms to the idea of freedom, providing opportunity for the other person to ask additional questions. As far as practical affairs are concerned, the spirit of Pfeiffer's work is to be found in what he had done. The spirituality from which his research springs can be discovered when one considers the context of his actions and the utility his results brought for life.

However, more needs to be said. Pfeiffer responded with utmost care to spiritual search whenever such effort made itself noticeable or whenever he could promote a vocational pursuit. In such instances he wrote detailed letters, pointing out possibilities, reflecting on all the details. In everyday life he might at times have shown a harsh exterior or appeared to be presumptuous. This is not important. Whenever one was in need of orientation or encouragement in a spiritual search, then Pfeiffer demonstrated caution, prudence, one can say even great tenderness. When he spoke, one felt set free. This kind of caution is a must, when questions of the inner path are at stake. Pfeiffer's gift to his fellow human beings was a practical objectivity in an air of liberating spiritual encouragement. (It goes without saying that the above remarks are observations of one individual and, hence, subjective. Others might well be able to add yet other perspectives.)

Pfeiffer's devotion to agriculture, to improving nutrition, and to methods of experimentally studying nature, springs from his inner spiritual life, from a strictly meditative practice, and his schooling by Rudolf Steiner whose spiritual science he studied. (One has to be most thankful to Alla Selawry, whose 1987 biography of Pfeiffer gives a detailed account of this aspect of his life. Selawry's work is based on many years of correspondence, meetings, and faithful cooperation.) In a lecture which Pfeiffer gave in 1958 in Dornach, he spoke of some of the motives of his work. Some of the following thoughts are taken from this lecture.

> When in 1920 as a young man I moved to Dornach
> and did all that work as an apprentice, I was
> pondering a very specific question ... which I put
> to Rudolf Steiner. This was the thought: If it is the
> nature of physical energies (electricity, magnetism,
> and, later, atomic energy) to eventually lead to
> disintegration, decomposition, and to bring down
> our civilization — would it not then be necessary
> to discover and use in technology an energy of
> life, which carries in it the laws of formation and
> nurture? This could awaken in humans, according
> to their nature, such ideas, feelings, and impulses of
> will which will teach them to attend to that which
> promotes growth, synthesis, construction, harmony.
> Does this sort of energy exist? Dr Steiner replied
> that this energy is active in the etheric body of man,
> in the etheric formative forces, in the warmth and
> light ether, the chemical and life ether. He pointed
> out that such energy could be studied in certain
> transformations of substances in the human body.

It was my idea, that by discovering and utilizing such etheric energy, a positive technology and social order, *proceeding from the Goetheanum, could be* carried out into the world. On Dr Steiner's suggestion, I did a few simple experiments. But he then saw that this was not yet the time to make known such energy. So he asked me to keep quiet. When I asked under what conditions it would be appropriate to discover and work with such energy, he replied that such energy must not be abused; hence, a social order along the lines of threefold order, at least in some small corner of the earth, would be needed. A second condition is that there be a general dissemination of Waldorf pedagogy, because such education makes it possible for children to enter into a sound spiritual development.

Healthy nutrition produced by biodynamic agriculture provides a physiological basis for this sort of positive development. Dr Steiner pointed out that, first the problem of nutrition must be solved before it will be possible for man to work in a spiritual way with etheric impulses. A healthy, well-balanced nutrition will release forces in man for spiritual pursuits. Ethical impulses will be furthered, Dr Steiner explained, by releasing the light forces in man's air-warmth organization, the sound forces of the chemical ether in his fluid organization, and the life forces in his solid organization. The food of our time does not contain these forces. Rather, it generates theoretical thought, with all its coldness regarding man's warmth organization, its paralysis of his light-air organization, and its deadening of his sound and life.

These are the tasks set for us, then. On the one hand, the aim is to prepare for the etheric forces which can prevent catastrophe of technology. On the other hand, it is the aim of every man to be active inwardly to find

the spirit in matter, to strengthen the forces of the heart,
etherealize his blood, and thus transform his body so
that it becomes spiritual and overcomes the lower nature
of matter. This is possible. It is possible to travel the path
of knowledge, not just sporadically, but continuously and
zealously.

Pfeiffer's life and work makes visible, how out of modern
consciousness, a person can summon his strength to promote
the life-giving principles of a spiritual world view in opposi-
tion to intellectual and materialistic one-sidedness of our time.
Considering the crisis in agriculture and our environment, the
threat to the life of the organism of the earth, such principles are
looked for in one form or the other. In keeping with his essential
spiritual impulse, Pfeiffer advocated such life-giving principles
in freedom, independent of government support. Regrettably,
although Pfeiffer had to devote much of his personal strength,
his valuable time, and his new ideas to earning the means for the
material existence, often his means were barely sufficient.

Here and there are notes which have been taken during the
anthroposophical lectures which Pfeiffer gave through the years
at Threefold Farm, in New York, and elsewhere. He participated
in the meetings of the Council of the Anthroposophical Society
in America. As far as can be ascertained, he gave extended series
of lectures on the cultural epochs, the evolution of consciousness,
and the spiritual powers which are active in this evolution. These
lectures related in particular to Rudolf Steiner's cycle on the spiritual guidance of man and mankind. These are activities which, on
the one hand, reflect Pfeiffer's love of observing nature and the life
in nature. On the other hand these activities also helped to foster
qualities of soul in those who met and heard Pfeiffer.

This mode of understanding nature is the same as spiritual
development into higher levels of the human being. The title of
the lecture from which the above remarks about etheric forces

were taken is *Subnature and Supernature in the Physiology of Plants and Man*. In that lecture, Pfeiffer dealt with the subject in the same way he had frequently before: he examined the specific results of scientific research in a comprehensive context of natural processes. Then step by step he revealed their essential nature. Finally, he encouraged responsible action. Putting before oneself the record of Pfeiffer's work and its results, one becomes aware that Pfeiffer's efforts stemmed from a powerful stream of spiritual endeavor.

Acknowledgment

From the mid 1950s until 1961, I was acquainted with Pfeiffer through gatherings and discussions in small, intimate groups. For a while I also shared with Pfeiffer an enjoyable working relationship on agricultural matters. Apart from personal contacts and Pfeiffer's printed works, the major source for this biographical sketch is Alla Selawry's deeply spiritual biography, *Ehrenfried Pfeiffer*, published in German in 1987. Most of the quotations given here are taken from her book. Further information was drawn from the reports on the early days of biodynamics in the United States which Evelyn Speiden-Gregg published in the journal *Bio-Dynamics* (1976/77).

Useful Contacts

If you would like to learn more about biodynamics, or the work of Ehrenfried Pfeiffer and Rudolf Steiner, please visit the website of the Biodynamic Association in North America at *www.biodynamics.com*. More information about biodynamics around the world can be found through local biodynamic associations.

Demeter International:
www.demeter.net
Australia:
Bio-Dynamic Research Institute
www.demeter.org.au
Biodynamic Agriculture Australia
www.biodynamics.net.au
Canada: Society for Bio-Dynamic Farming & Gardening in Ontario
biodynamics.on.ca
India: Bio-Dynamic Association of India (BDAI)
www.biodynamics.in
New Zealand: Biodynamic Farming and Gardening Association
www.biodynamic.org.nz
South Africa: Biodynamic Agricultural Association of Southern Africa
www.bdaasa.org.za
UK: Biodynamic Association
www.biodynamic.org.uk
USA: Biodynamic Association
www.biodynamics.com

Further reading

Karlsson, Britt and Per, *Biodynamic, Organic and Natural Winemaking*, Floris Books, UK.

Klett, Manfred, *Principles of Biodynamic Spray and Compost Preparations*, Floris Books, UK.

Koepf, H.H. *Koepf's Practical Biodynamics*, Floris Books, UK.

Kranich, Ernst Michael, *Planetary Influences upon Plants*, Biodynamic Association, USA.

Masson, Pierre, *The Biodynamic Manual*, Floris Books, UK.

Osthaus, Karl-Ernst, *The Biodynamic Farm*, Floris Books, UK.

Pfeiffer, Ehrenfried, *The Biodynamic Orchard Book*, Floris Books, UK.

—, *The Earth's Face*, Lanthorn Press, UK.

—, *Weeds and What They Tell Us*, Floris Books, UK.

Philbrick, Helen and Gregs, *Companion Plants: An A to Z for Gardeners and Farmers*, Floris Books, UK.

Sattler, F. & E. von Wistinghausen, *Growing Biodynamic Crops*, Floris Books, UK.

Steiner, Rudolf, *Agriculture (A Course of Eight Lectures)*, Biodynamic Association, USA.

—, *Agriculture: An Introductory Reader*, Steiner Press, UK.

—, *What is Biodynamics? A Way to Heal and Revitalize the Earth*, Steinerbooks, USA.

Storl, Wolf, *Culture and Horticulture*, North Atlantic Books, USA.

Thun, Maria, *Gardening for Life*, Hawthorn Press, UK.

—, *The Biodynamic Year*, Temple Lodge Publishing, UK.

Thun, Matthias, *The Maria Thun Biodynamic Calendar* (annual), Floris Books, UK.

—, *The North American Maria Thun Biodynamic Almanac* (annual), Floris Books, UK.

von Keyserlink, Adelbert Count, *The Birth of a New Agriculture*, Temple Lodge Publishing, UK.

—, *Developing Biodynamic Agriculture*, Temple Lodge Publishing, UK.

Weiler, Michael, *The Secrets of Bees*, Floris Books, UK.

Wright, Hilary, *Biodynamic Gardening for Health and Taste*, Floris, Books, UK.

You may also be interested in...

Weeds and What They Tell Us

Ehrenfried E. Pfeiffer

This wonderful little book covers everything you need to know about the types of plants known as weeds. Ehrenfried Pfeiffer discusses the different varieties of weeds, how they grow and what they can tell us about soil health.

The process of combatting weeds is discussed in principle as well as in practice, so that it can be applied to any situation.

The Maria Thun
Biodynamic Calendar

This useful guide shows the optimum days for sowing, pruning and harvesting various plants and crops, as well as working with bees. It is presented in colour with clear symbols and explanations.

The calendar includes a pullout wallchart which can be pinned up in a barn, shed or greenhouse as a handy quick reference.

Have you tried our Biodynamic Gardening Calendar app?

A quick, easy way to look up the key sowing and planting information found in the original *Maria Thun Biodynamic Calendar*.

- 🌼 Filter activities by the time types of the crops you're growing
- 🌼 Automatically adjusts to your time zone
- 🌼 Plan ahead by day, week or month
- 🌼 Available in English, German and Dutch

florisbooks.co.uk

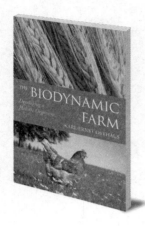

The Biodynamic Farm

Developing a Holistic Organism

Karl-Ernst Osthaus

'*This is more than just a book for the farmer, but rather one for anyone who would like to understand and work with Nature.*'

– SCIENTIFIC & MEDICAL NETWORK REVIEW

Large-scale agriculture tends to view a farm as a means for producing a certain amount of grain, milk or meat. This practical book argues instead for a holistic method of farming: the farm as a living organism. This is the principle of biodynamic farming.

This is an invaluable book for anyone considering setting up a farm, or developing their existing farm with new biodynamic methods.

florisbooks.co.uk

Koepf's Practical Biodynamics

Soil, Compost, Sprays and Food Quality

Herbert H. Koepf

'Books like this remind us of the parallels between plant and human health, and also all the rhythms of life as a process in time regulated by both the earthly and cosmic environments.'

– SCIENTIFIC & MEDICAL NETWORK REVIEW

Herbert Koepf was a pioneer of biodynamics in Germany, the USA and in the UK. He was an expert teacher, and drew on his own practical background in farming.

This is an invaluable guide for anyone working with biodynamic methods, offering Koepf's unique insights and wisdom on practical issues.

florisbooks.co.uk

Growing Biodynamic Crops

Sowing, Cultivation and Rotation

Friedrich Sattler &
Eckhard von Wistinghausen

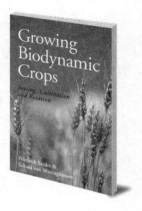

A biodynamic farm is an integrated, holistic organism which balances animal husbandry with growing a range of plants, crops and trees. Balance is of the utmost importance, and will result in a sustainable farm.

This book focuses in depth on one aspect of biodynamic farming: growing crops. It addresses all aspects of crop husbandry, from the nature of plants and issues of land use to cultivating grassland, weed control, crop rotation, seeds and sowing, and growing cereals, row crops, legumes, fodder crops and herbs.

This is a comprehensive overview of crops and cropping for biodynamic farmers, written by experts in their field.

florisbooks.co.uk

Biodynamic Beekeeping

A Sustainable Way to Keep Happy, Healthy Bees

Matthias Thun

Biodynamic Beekeeping is the first book to offer practical instruction on caring for bees using biodynamic theories and methods. By considering the influence of the movement of the stars and planets on the bees' natural habits, biodynamics encourages beekeepers to be more in tune with their bees indicating, for example, the best days on which to inspect colonies or gather honey.

This fascinating book offers beekeepers detailed advice on how to work more holistically including the challenges and advantages of breeding queen bees.

Also available as an eBook

florisbooks.co.uk

More essential reading for biodynamic growers

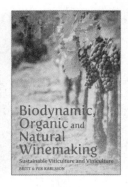

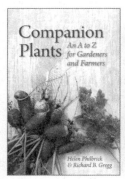

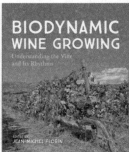

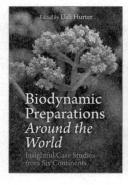

florisbooks.co.uk

Floris
Books

For news on all our **latest books,**
and to receive **exclusive discounts,**
join our mailing list at:

florisbooks.co.uk/signup

Plus subscribers get a FREE book
with every online order!